DAVIS'S

MANUAL OF MAGNETISM,

INCLUDING

GALVANISM, MAGNETISM, ELECTRO-MAGNETISM,
ELECTRO-DYNAMICS, MAGNETO-ELECTRICITY,
AND THERMO-ELECTRICITY.

WITH 180 ORIGINAL ILLUSTRATIONS.

THIRTEENTH EDITION.

BOSTON:
PUBLISHED BY THOMAS HALL,
(SUCCESSOR TO DANIEL DAVIS,)
MAGNETICAL AND TELEGRAPHIC INSTRUMENT MAKER,
13 BROMFIELD STREET.
1865.

The Instruments described in the following treatise are manufactured by Thomas Hall, and sold by him, at No. 13 Bromfield Street, Boston, Mass.

NOTICE.

THIRTEENTH EDITION.

Owing to the great advance in the price of Stock, Workmen, Materials, &c., the subscriber is obliged to increase his prices thirty per cent. above the prices stated in the following Catalogue.

Up to the present time we have sold at the prices enumerated in the Catalogue, but find it impossible to continue manufacturing without the above increase in our prices, as every thing used in connection with the business is higher by thirty per cent., and in many instances fifty per cent., than formerly.

We trust that our patrons will see the necessity for this increase, and govern themselves accordingly.

THOMAS HALL.

Boston, January 1, 1865.

Wright & Potter, Printers, 4 Spring Lane.

PREFACE.

THE present wcrk is principally occupied with Magnetism in its connection with Electricity. But the general phenomena of both these sciences are described as fully as the thorough comprehension of the relations existing between them appeared to require. It has been the aim to give to the book a practical, rather than a theoretical character, and to introduce hypotheses no further than was essential to a clear explanation of the phenomena described. Many important observations and results connected with magnetism and electricity have never been brought together in any scientific work; and the present volume, though mainly intended as a companion to the apparatus manufactured by me, has, in consequence, taken the form of a treatise on those branches of science to which it relates, and may be used as a text-book.

Since the appearance of the first edition of this Manual, in 1842, some progress has been made in the departments of science treated of, and many new and improved instruments have been contrived to illustrate them. In preparing a new edition, alterations and additions have been made, to adapt it better to the purposes of a text-book for Colleges and High Schools, and also as a companion to the apparatus. The Manual has for some years been in use as a text-book, for the sake of its demonstrations, in the United States' Military Academy at West Point.

The preparation of the present, as well as of the former edition, has been principally under the charge of Drs. John Bacon, Jr., and William F. Channing. The arrangement which has been adopted, and which differs from that of any previous work, is founded upon a natural order; and the attempt has thus been made to establish a more scientific relation between the facts of magnetism. Many of the instruments and experiments are original, and are now for the first time described. Among the new articles of apparatus, a considerable number have been contrived or improved by Mr. Thomas Hall. Numerous wood cuts are introduced to illustrate the subjects under consideration.

DANIEL DAVIS.

BOSTON, *January* 1, 1865.

CONTENTS.

INTRODUCTION.

PRODUCTION OF ELECTRICITY.

MAGNETISM.

BOOK I.

DIRECTIVE TENDENCY OF THE MAGNET.

BOOK II.

INDUCTION OF MAGNETISM.

BOOK III.

INDUCTION OF ELECTRICITY.

INTRODUCTION.

DEFINITIONS AND EXPLANATIONS.

1. Magnetism. — The term *magnetism* expresses the peculiar properties of attraction, repulsion, &c., possessed, under certain circumstances, by iron and some of its compounds, and in a somewhat inferior degree by nickel, a closely-allied metal. Cobalt is perhaps slightly magnetic.

Electro-Magnetism. — That branch of science which relates to the development of magnetism by means of a current of electricity, is called *electro-magnetism*. It will be treated of in Book I. Chapter 2, and in Book II. Chapter 2.

Magneto-Electricity treats of the development of electricity by the influence of magnetism, and will form the subject of Book III. Chapter 2.

2. Magnet. — The body which exhibits magnetic properties is called a *magnet*. This name is confined to the metallic substances mentioned above; but all conductors of electricity are capable of exhibiting similar attractions and repulsions while conveying a current.

NATURAL MAGNETS. — Certain ores of iron are found to be possessed of the magnetic properties in their natural state. These are called *natural magnets*, or *loadstones.*

ARTIFICIAL MAGNETS. — Bodies belonging to the magnetic class, in which magnetism is artificially *induced*, are called *artificial magnets.*

3. INDUCTION OF MAGNETISM. — Whenever magnetic properties are developed in bodies not previously possessed of them, the process is termed the *induction of magnetism.* When this is effected by the influence of a magnet, it is called *magnetic induction;* when by a current of electricity, *electro-magnetic induction.*

INDUCTION OF ELECTRICITY. — This term expresses the development of electricity by the influence of other electricity in its neighborhood, or by the influence of magnetism. In order to distinguish the inductive action of an electric current from the *static induction* of electricity at rest, the former is called *electro-dynamic induction.* The development of electricity by the influence of a magnet is termed *magneto-electric induction.*

4. POLES. — The magnetic phenomena manifest themselves principally at the two opposite extremities of the magnet; the force of the attractions and repulsions diminishing rapidly as the distance from them increases, until it becomes entirely insensible at the middle point. These extremities are called the *poles* of the magnet.

5. The earth itself is found to possess the properties of a magnet, having magnetic po.es corresponding nearly in their direction with the poles of its diurnal rotation. Now, if a straight magnet be suspended so as to allow of a free horizontal motion, it will be found to place itself in a direction nearly north and south; as will be explained hereafter. The end which turns towards the north is called the *north pole* of the magnet, the other end its *south pole*. Hence every magnet, whatever its form, is said to have a north and a south pole. In the figures to be hereafter described, the north pole is indicated by the point of an arrow, and the south pole by the feather; or by the letters N and S respectively. The *poles* of a galvanic battery will be described farther on, when treating of that instrument.

6. Permanent Magnets.—It is found that pure soft iron easily acquires magnetism when exposed to any magnetic influence, but immediately loses this magnetism when that influence is withdrawn. But steel, which is a compound of iron with a small quantity of carbon, and especially hardened cast-steel, though it acquires the magnetic properties less readily, retains them more or less permanently after they are acquired. Hence a magnet formed of hardened steel is called a *permanent magnet*.

7. Bar Magnet.—An artificial permanent magnet, in the form of a straight bar, is called a *bar magnet*.

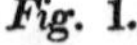

Fig. 1.

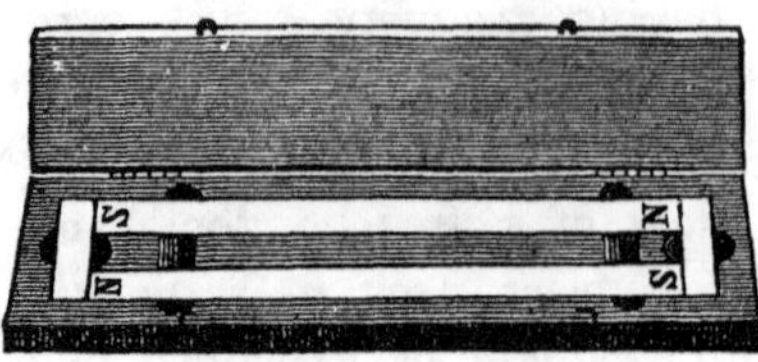

Fig. 1 represents a small case, containing two bar magnets, with two short pieces of soft iron connecting their poles: these act as *armatures* (see § 9), and serve to preserve the power of the magnets. The magnets, when not in use, should be kept packed in the case, with their opposite poles connected by the armatures, in the manner shown in the cut.

Compound Bar Magnet. — A magnet composed of several straight bars joined together, side by side, with their similar poles in contact, for the purpose of increasing the magnetic power, is called a *compound bar magnet.*

Fig. 2.

Such a magnet, composed of three simple magnets, fastened together, is represented in Fig. 2.

Fig. 3.

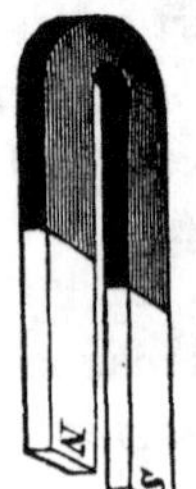

8. Horseshoe or U-Magnet. — A magnet which is bent into such a form as to bring the two opposite poles near together, so that they can be connected by a short, straight piece of iron, is called a *horseshoe* or *U-magnet.*

Fig. 3 represents a steel magnet of this description.

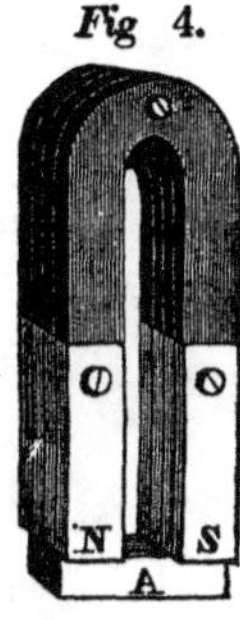

Fig 4.

Compound Horseshoe Magnet. — A magnet composed of several horseshoe magnets joined together, side by side, as in Fig. 4, for the purpose of increasing the power, is called a *compound horseshoe magnet*, or *magnetic battery.* These magnets are charged separately, and are put together with all the similar poles in the same direction.

9. Armature. — A piece of soft iron, adapted to, and intended to connect the poles of a magnet, is called an *armature*, or *keeper.* Horseshoe magnets are usually provided with an armature, consisting of a straight bar of iron, for the purpose of preserving their magnetic power: this should be kept constantly applied to the poles of the magnet when it is not in use; as shown in Fig. 4, where A is the keeper. Armatures are employed in various experiments, and their forms vary with the purposes intended.

Fig 5.

10. Magnetic Needle. — A light and slender magnet, mounted upon a centre of motion, so as to allow it to traverse freely in certain directions, is called a *magnetic needle.* It may be so mounted as to move only horizontally, as in Fig. 5; or its motion

Fig. 6.

may be confined to a vertical direction, as shown in Fig. 6. This last form is called a *dipping needle.* By means of a universal joint, the needle may be supported in such a manner as to have freedom of motion in both a horizontal and vertical direction.

11. The most obvious effects exhibited by magnets are their power to attract iron, and their tendency, when freely suspended, to assume a determinate position in reference to the earth. For a long time, these were the only properties which were noticed, or at least which received particular attention. The attractive power of the loadstone over small pieces of iron seems to have been known from the remotest antiquity; but its polarity with regard to the earth does not appear to have been observed, at least in Europe, until the eleventh or twelfth century of the Christian era. It is, however, stated, by a distinguished orientalist, M. Klaproth, that the Chinese were acquainted with the polarity of the magnet, and employed it as a guide to travellers on land, at least as early as the second century also, that the use of the mariner's compass in navigation originated with them, and was communicated to the European nations through the Arabs.

PRODUCTION OF ELECTRICITY.

12. As a current of electricity is requisite in many of the experiments to be mentioned hereafter, it becomes necessary to describe the various means by which it may be produced.

I. MECHANICAL OR FRICTIONAL ELECTRICITY.

13. The electricity developed by the electrical machine is called *mechanical* or *frictional*, from the mechanical force or friction by which it is obtained. It possesses properties differing much in degree from those exhibited by the galvanic arrangements described below, and is far inferior in producing magnetic effects, which require an *electric current.* On the other hand, it greatly surpasses galvanic electricity in exhibiting the phenomena of electricity at rest, under its two forms of *positive* and *negative.*

14. The great development of electricity observed during the escape of steam from high pressure boilers, may be mentioned here. This is collected for purposes of experiment, by plunging into the steam, escaping from the safety valve, a brass rod (Fig. 7) furnished with a brush of points, P, at one end, to collect the electricity, and held by means of an in-

Fig. 7.

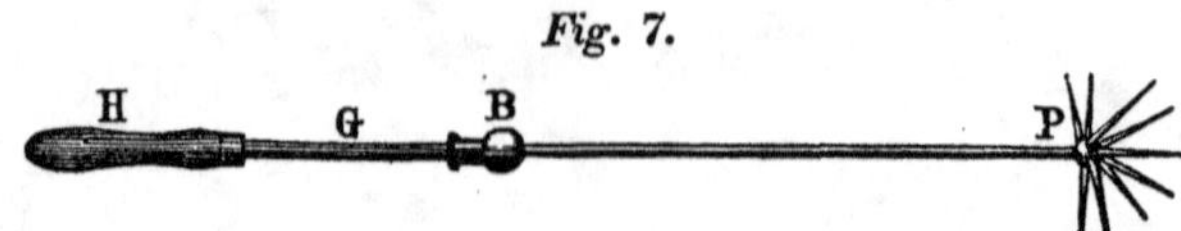

sulating handle attached to the other end. A length of six or eight feet is found advantageous, in this instrument, to convey and insulate the electricity, which may be conveniently drawn from the lower part of the rod. In the cut, the brass rod is represented as terminating in a brass ball, B, insulated from the wooden handle, H, by a stout glass rod, G.

15. The electricity obtained in this way from steam is of high intensity, affording sparks of an inch or more in length, and charging the Leyden jar so as to give strong shocks. It is almost always *positive,* and is not obtained unless the steam is of high pressure, so as to issue from the valve as a transparent vapor.

II. GALVANIC OR VOLTAIC ELECTRICITY.

16. These names are given to that form of electricity which is produced by chemical action. It is found that, when two metals are placed in contact with each other, and with some liquid capable of acting upon one more than upon the other, electricity of a peculiar character is developed. The peculiar electrical relation of the metals employed, also exerts an influence upon this result. The metals most

extensively used are zinc and copper, or zinc and platinum; and the chemical agent is some liquid containing an acid having a powerful affinity for zinc. The language adopted in describing the resulting phenomena was founded originally on the supposition that electricity is given out to the copper from the zinc, which is corroded, through the liquid between them. This is shown in the adjoining cut (Fig. 8), which represents a glass vessel, nearly filled with the fluid, and containing a zinc plate, marked Z, and one of copper, C. Now, the supposed motion of the electric current within the vessel, is from Z to C; and if the two metals are connected by a wire without the vessel, as in the cut, in order to fulfil the condition of metallic contact, the electricity is supposed to pass around through the wires from the copper to the zinc again, to restore the equilibrium of the fluid. Thus the current is considered as passing from zinc to copper *within* the series, and from copper to zinc *without* it. The wire connected with C is called the *positive* pole of the arrangement, and that with Z the *negative* pole.

Fig. 8.

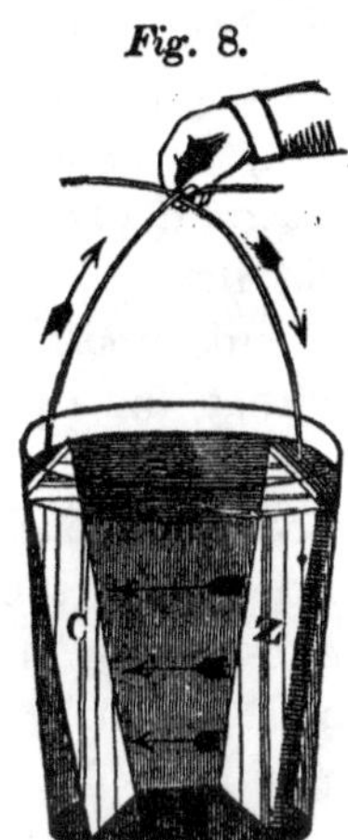

17. The electricity proceeding from the positive pole is the same in its relations as the electricity from the prime conductor of the electrical machine, which originally received the name of *positive;* while

that from the negative pole corresponds with the electricity obtained from the rubber of the machine. These terms are, however, to a certain extent, arbitrary. It is still an open question, whether there is one fluid moving in a particular direction, or two fluids moving in opposite directions, or no motion of a fluid at all. The fact which is sought to be explained by these theories remains fixed. In the above-described circulation of the electrical fluid, technically called the *galvanic circuit,* there is an electrical influence propagated in a certain unchanging direction; and as the control of the magnetic and chemical reactions produced depends upon our knowledge of this, it is necessary that the signification of the terms should be understood.

18. Professor Faraday has proposed a nomenclature of electricity, which has been adopted in some scientific treatises. The poles are called by him *electrodes,* from the Greek ἤλεκτρον and ὁδός; that is, ways or paths of electricity; the positive pole the *anode*, from the Greek ἄνοδος, an ascending or entering way, and the negative pole the *cathode,* from the Greek κάθοδος, a descending way or path of exit. The terms *positive* and *negative pole* are, however still most frequently employed to designate these extremities, and the wire without, when in connection with these poles, is spoken of as the channel of a positive current passing from the former to the latter.

19. Instead of using two metals to form the galvanic circuit, one metal, in different states, may

be used on the same principle; the essential condition of this current being only that one part of a conductor of electricity shall be more corroded by some chemical agent than another part. Thus, if a galvanic pair be made of the same metal, one part of which shall be softer than another, as of cast and rolled zinc, so as to be differently corroded, or if a greater amount of surface be exposed to corrosion on one side than on the other, or a more powerful chemical agent be used on one side, a current will be determined from the part most corroded through the liquid to the part least corroded, whenever the circuit of the poles is completed.

20. Galvanic electricity is capable of producing the most extensive magnetic, chemical, and calorific effects. In this respect, it has a far greater capacity than mechanical electricity, though it is found that, by the accumulation of this latter, the same effects can be produced in proportion to the amount present. This has led to the natural inference, that, in galvanic electricity, the *quantity* present is immense, while in mechanical, the *quantity* is small. On the other hand, it is found that, in the latter form, the electricity is much more energetic in its physical reactions; that it appears to be condensed upon insulated matter, and strives to obtain an equilibrium by diffusion in every direction. It is, therefore, said that mechanical electricity has more *intensity* than galvanic, though it is difficult to assign other than a general idea to this word. Owing to this difference of intensity, the substances, such

as glass, earthen ware, wood, ivory, which act as non-conductors to galvanic electricity, are much more numerous than the corresponding class, with reference to mechanical electricity. A true comparison between these two forms of electricity, would be made between the galvanic current and a current of mechanical electricity, freely circulating, as through a wire connecting the prime conductor and the rubber of an electrical machine. So compared, the fluid would be found in both identical in its effects, with scarcely greater difference in the conditions of quantity and intensity than we are able to produce by different arrangements of galvanic series.

21. There are two modes by which the peculiar powers of a galvanic arrangement may be increased — first, by increasing the size of the plates used; and, secondly, by increasing their number. 1. *The extension of the size of the plates.* If the size of the plates, that is, the extent of the surfaces acted upon by the chemical agent, is increased, some of the resulting effects become more powerful in the same ratio, while others do not. The power to develop heat and magnetism is increased, while the power to decompose chemical compounds and to affect the animal system is very slightly, or not at all, augmented. Batteries constructed in this way, of large plates, are sometimes called *calorimotors*, from their great power of producing heat; and they usually consist of from one to eight pairs of plates. They are made of various forms. Sometimes the sheets

of copper and zinc are coiled in concentric spirals, sometimes placed side by side; and they may be divided into a great number of small plates, provided that all the zinc plates are connected together, and all the copper plates together, and then, finally, tnat the experiments are performed in a channel of electrical communication opened between the one congeries and the other; for it is immaterial whether one large surface be used, or many small surfaces, electrically connected together. The effect of all these arrangements, by which the metallic surface of a single pair is augmented, is to increase the *quantity* of the electricity produced.—2. *The extension of the number of the plates consecutively*; that is, by connecting the copper plate of each pair with the zinc plate of the next pair. By this arrangement, the electricity is obliged to traverse a longer or shorter series of pairs; each pair being separated from the adjoining ones by a stratum of imperfectly conducting liquid, or by the walls of an insulating cell. The result is, that the electricity acquires that additional impulse, which has already been referred to, as *intensity*. It has greater power to pass through imperfect conductors, or through intervals in the circuit, to give shocks to the animal system, and to decompose chemical compounds; and when the number of consecutive pairs of plates is increased to some thousands, or even hundreds, the electricity developed approaches very near in its character to that produced by the electrical machine; it manifests similar attractions and repulsions, and in fact the

Leyden jar may be charged with it. With a very extensive series excited by water only, and in which each cell was carefully insulated, an English electrician has lately obtained an electrical discharge between the poles, although separated to a considerable distance. The electricity from one pair of plates has a very low intensity. As the number of consecutive pairs is multiplied, the intensity increases, until at length it approximates to that of frictional electricity, which is able to strike across a considerable interval of air, and to fracture solid nonconductors interposed in its circuit.

22. In consequence of the low intensity of the electricity required for electro-magnetic experiments, it is very easy of insulation. This is a great advantage in regard to the practical construction of magnetic apparatus. Where electricity exists in a state of high intensity, it has a strong tendency to pass off and dissipate itself through imperfect conductors; but where it exists only in great *quantity*, it requires nearly perfect conductors to allow it a passage. The electricity developed by a single pair of plates, however much its power may be increased by increasing the size of the plates, will scarcely pass across the smallest interval of air; and a wire conveying the current may be perfectly insulated by a covering of varnish. In working the electrical machine, on the other hand, the electrified parts of the apparatus must be kept at a distance from each other, raised on glass supports, or suspended by silken lines; and then, unless the atmosphere is very dry, the

electricity will be rapidly dissipated. But in the case of currents of low intensity, however great what is called the *quantity* may be, two wires may lie side by side, with a coating of varnish or wax between them, and convey different and opposite currents, without any perceptible electrical intercommunication.

23. For the purposes of electro-magnetic experiments, electricity of a low intensity is required; the power of the magnetic effects of a current of electricity depending upon an increase of its *quantity*, mainly. Increasing the number of consecutive pairs would only add to the *intensity* of the current, making it more unmanageable in respect to insulation, without adding much to its magnetic effects. Galvanic batteries having many pairs of plates are therefore unsuitable for these experiments. The maximum magnetic effect is produced by a single galvanic combination, or at most by five or six. In some of the reactions, however, of a magnet with a coil or electric current, as well as in experiments on decomposition and deflagration, batteries capable of producing intense currents are required. The most convenient forms of single pairs and of series will therefore now be described.

24. SMEE'S BATTERY. — This form will be first spoken of, as furnishing an example of the galvanic battery in its most simple form, as it has been already illustrated in Fig. 8. It consists of a glass tumbler, or other receiving vessel, on which a little frame rests, supporting the apparatus within. The metals employed

Fig. 9.

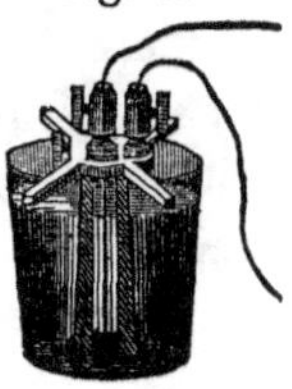

are platinum and zinc. The original battery of Smee consisted of zinc, and silver covered with a film of platinum; but the metal platinum itself is found so much superior to the platinized silver, and the difference in expense so slight, that it has been substituted. The arrangement, also, as shown in the figure, has been modified from the original form. The metal platinum is used in this case, as it is the least oxidizable of the metals, and therefore capable of producing a more powerful current with zinc than any other. On the frame will be seen two screw-cups for the attachment of wires by which the current may be conveyed in any direction. One of these screw-cups communicates with a strip of platinum foil, which is suspended between two zinc plates, both of the surfaces of the platinum being opposed to the zinc. The amount of galvanic action is generally in proportion to the metallic surfaces of different kinds opposed to each other, and also in proportion to the nearness of those surfaces. The other screw-cup is connected with both the zinc plates, thus uniting them into a single element of the pair. The screw-cup connected with the platinum is insulated from the metallic frame which supports it, by rosewood; and a thumb-screw, seen at the left, confines or releases the zinc plates, so that they can be renewed from time to time.

25. The liquid used to excite this battery is sulphuric acid (oil of vitriol), diluted with ten or twelve parts of water by measure. This acid acts on common zinc when the galvanic circuit is not estab-

lished, and a great loss would therefore ensue from the corrosion of the plates during the interval of experiments, even if the zinc was withdrawn from the acid whenever practicable. To remedy this defect, it has been found that zinc which has its surface amalgamated (or combined with mercury) withstands the action of diluted sulphuric acid, unless t is in galvanic connection with another metal; and accordingly the zinc of this battery, and of other batteries in which acid is used, is commonly amalgamated. It then remains almost uncorroded until the galvanic circuit is completed by making contact between the wires, when the zinc is immediately attacked. The amalgamation of the zinc is easily effected by rubbing it with a little mercury and muriatic acid at the same time. This battery, when once in action, is very constant. It does not, however, like the batteries hereafter to be described, arrive instantly at its highest rate of action when the circuit is completed, but takes an appreciable time to reach this point, and it is not therefore fitted for use with apparatus where the circuit is rapidly broken and renewed. No adequate increase of power is obtained by adding to the size of this battery.

26. In order to understand some of the phenomena which will be spoken of hereafter, it is necessary to notice what takes place when the circuit of the battery is completed and the galvanic current begins to flow. The first thing which is observed is a rapid evolution of gas in bubbles from the platinum plate. It was stated that electricity

was supposed to pass through the liquid between the plates in a direction from the zinc to the copper or platinum. This passage of electricity through fluids which are not themselves elementary, is attended always, according to Faraday, with decomposition, whether the fluid is between the plates of the battery or interposed in the course of the wires or poles leading from it. Thus, where a battery is excited simply by water, that fluid is decomposed; its oxygen attacks the zinc, and its hydrogen is given off in contact with the electro-negative metal.

27. In Smee's battery, the water of the acid solution may also be considered as undergoing decomposition, one atom of its oxygen uniting with one atom of zinc, in order to enable the sulphuric acid to unite with the resulting oxide of zinc, and the corresponding atom of hydrogen being given off in contact with the platinum. It will be observed that the oxygen and hydrogen appear at the opposite sides of the fluid undergoing decomposition. It is not, however, believed that these elements travel through the intervening distance, but that the two atoms of water in contact with the plates are simultaneously decomposed, and that a chemical equilibrium is then established by a progressive exchange of elements in all the intervening particles. The subject of electro-decomposition will be further spoken of in connection with the electrotype.

28. The gas which is so plentifully disengaged at the surface of the platinum in this battery is apt to bring the strip into contact with the zinc, and thus

cause a discharge between the metals within the battery, instead of through the poles without. To obviate this, the platinum is either confined in its proper position by some fixture at the bottom, or a bead of glass is attached to it, which prevents its swinging against the zinc.

Fig. 10.

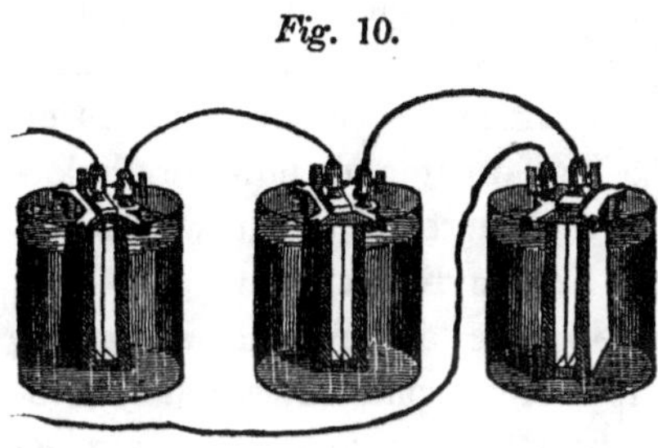

Fig. 10 represents a series of three pairs of this battery, in which it will be observed that the platinum of one is connected with the zinc of the next, and that the terminal wires proceed, consequently, one from a platinum plate and the other from a zinc plate, as in a single pair.

29. Cylindrical Sulphate of Copper Battery. — This battery, a vertical section of which is represented in Fig. 11, consists of a double cylinder of copper, C C, with a bottom of the same metal; which answers the purpose both of a galvanic plate and of a vessel to contain the exciting solution. The space between the two copper cylinders receives the solution. There

Fig. 11.

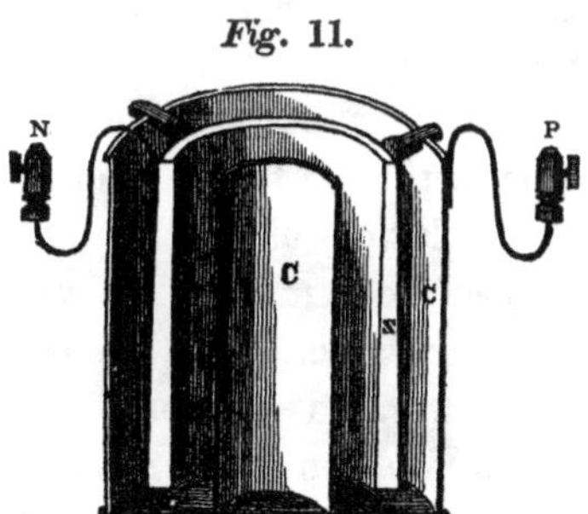

is a movable cylinder of zinc, marked Z in the sectional view, which is let down into this solution whenever the battery is to be put in action. It is, of course, intermediate in size, as well as in position, to the two copper cylinders, and is made to rest upon the exterior one by means of three insulating branches of wood or ivory, projecting from it outwardly. Thus it hangs suspended in the solution, and presents its two opposite surfaces to the action of the liquid, and to the inner and outer cylinders of copper respectively. There is a screw-cup, N, connected with the zinc cylinder, and also one marked P, with the copper cylinder; and, according to the principles heretofore explained, when a communication is made between these cups, the electricity developed by the action within the battery will pass from one to the other.

Fig. 12.

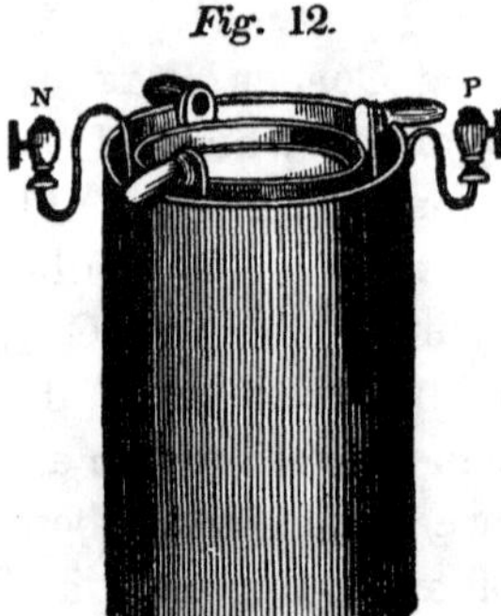

Fig. 12 is a perspective view of the same battery, in which the parts will be understood without further explanation.

30. The liquid employed to put this battery in action is a solution of sulphate of copper (common blue vitriol) in water. To prepare it, a saturated solution of the salt is first made, and to this solution is then added as much more water. It may be convenient to know, that a pint of water, at the ordinary temperature of

the atmosphere, is capable of dissolving one fourth of a pound of blue vitriol; so that the half-saturated solution employed will contain about 2 oz. of the salt to the pint. The addition of a small portion of alcohol to this solution is sometimes of advantage, by increasing the permanence of its action.

31. The action in this case is as follows: The zinc is oxidized by the oxygen of the water; the oxide combines with the acid of the salt, forming sulphate of zinc, which remains in solution; while the oxide of copper, which was previously combined with the acid, being set free, partly adheres to the surface of the zinc cylinder, or falls to the bottom of the solution as a black powder, and partly is reduced to metallic copper, which is precipitated on the surface of the copper cylinder, or falls to the bottom in fine grains. This reduction of the oxide to the metallic state takes place in the following manner: The water of the solution furnishes oxygen to the zinc, and thus enables it to combine with the acid; while the hydrogen, which is liberated, again forms water with the oxygen of the oxide of copper with which it comes in contact, leaving the metal free. Hence but little gas is given off during the action of a battery charged by sulphate of copper, as the hydrogen, which usually escapes, is in this case mostly absorbed.

32. In this form of battery, there is greater intensity in proportion to the quantity of electricity produced than in Smee's, from the fact that the final decomposition falls on oxide of copper, and not on

water, which is separated into its elements with greater difficulty than the former. The force of a galvanic battery is equal, other conditions being the same, to the amount of corrosive force in the solution, minus the force of chemical affinity which has to be overcome in the decompositions which necessarily attend the process. The sulphate of copper battery is found more convenient than any other in a large class of experiments, and also generally in the hands of persons to whom the use of acids would be inconvenient. The solution of sulphate of copper is entirely neutral, and does not injure the colors or texture of organic substances. Smee's battery is a more condensed form, and the action on an equal surface of zinc more energetic.

33. The coating of oxide of copper should always be removed from the zinc after using the battery. For this purpose a card brush is provided. With this the surface of the zinc should be thoroughly cleansed, with the aid of plenty of water, whenever it has been in use. If this has been neglected, so that the zinc has become covered, in whole or in part, with a hard coating, it will be necessary to scrape or file it, to obtain a clean metallic surface. The deposit of copper, also, which will gradually accumulate below, must be removed from time to time.

34. Batteries excited by sulphate of copper will keep in good action for twenty or thirty minutes at a time. The zinc cylinder should always be taken out of the solution when the battery is not in use, but the solution itself may remain in the battery, as it has no

chemical action upon the copper, but tends to keep its surface in good condition. When the solution has lost its power, as it will do, of course, after a time, it is not best to attempt to renew its efficiency by adding a fresh quantity of the salt. It should be thrown away, and a new solution be prepared, according to the foregoing directions.

35. Fig. 13 represents a PROTECTED SULPHATE OF COPPER BATTERY. — It has been stated, in § 31, that the zinc soon becomes coated with oxide of copper or pulverulent metallic copper, and the action of the battery consequently declines. Now, the efficiency of sulphate of copper in providing oxide of copper to undergo decomposition in place of water, is equally great in the final result when it forms only that part of the solution next the copper plate. Accordingly, by interposing a cell of porous earthen ware between the zinc and copper, two cells are produced, in the outer one of which sulphate of copper may be used, and in the inner one a solution of Glauber's salt (sulphate of soda), or common salt. These do not coat the zinc, though sufficient of the copper solution will eventually pass through to make it necessary occasionally to clean the zinc plate. Dilute sulphuric acid may also be used in the inner cell; but in this case, it is necessary or advisable to amalgamate the zinc. A lip will be observed in the figure, which is used for adding

Fig. 13.

crystals of sulphate of copper to the solution as its strength declines.

36. Fig. 14 shows the internal construction of this form. In the centre is seen a solid rod of zinc,

Fig. 14.

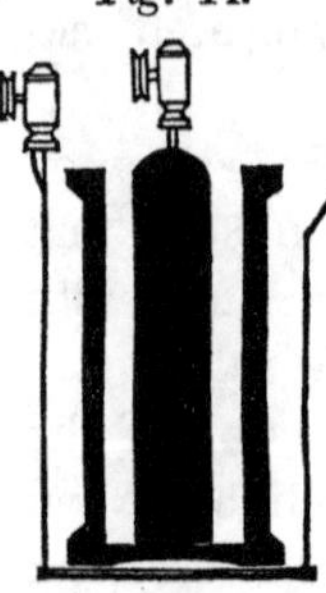

with the space for the saline or acid solution surrounding it. Next comes the porous cell; and without this again the space containing the sulphate of copper solution. It is necessary, after the porous cell has been in use, to allow it to soak in several waters before drying it to put away. Otherwise the salts which are contained in it will crystallize in the interstices and disintegrate its substance.

Fig. 15.

37. Another form of protected sulphate of copper battery is represented in Fig. 15, the chief difference consisting in the porous cell, which is of thick leather. Except, however, where batteries of a particular shape or size are required, for which earthen ware cells cannot readily be obtained, those previously described are preferred.

Fig. 16.

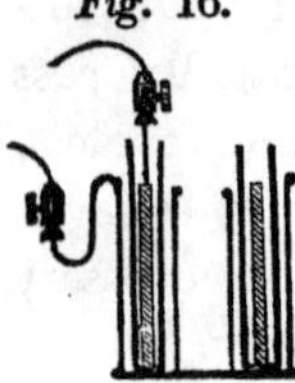

38. Fig. 16 represents a protected battery, in which the leather forms a double cell, enclosing the zinc, so that a double cylinder of copper can be used, as in the cylinder battery (Fig. 11), and both surfaces of the zinc are thus opposed to corresponding surfaces of copper

Other porous or membranous substances, such as thick brown paper, or bladder, will answer the same purpose as leather, in preventing the ready admixture of the solutions, and allowing a free passage to the electrical current. When the partition is sufficiently thin or permeable, the battery is as powerful as if charged in the usual mode with a solution of blue vitriol.

39. The action within this class of batteries is as follows: The zinc is oxidized as usual, at the expense of the water of the solution which surrounds it; while the hydrogen, instead of being given off at the surface of the negative plate, as in most batteries, decomposes the sulphate of copper, forming water with the oxygen of the oxide of copper, and liberating the sulphuric acid, which passes through the porous partition into the other cell. A gradual and steady supply of acid is thus furnished to dissolve the constantly forming oxide of zinc.

40. The protected battery is also commonly called *constant*, or *sustaining*, from the fact that it will maintain a nearly unvarying power for several days in succession, if the solution of sulphate of copper is kept saturated by occasionally adding a little of the pulverized salt and stirring the liquid, to make it of uniform strength. With weaker solutions, or a less permeable partition, an action of sufficient energy for many purposes may be sustained for a week or more. and, when it declines, may be renewed by cleaning the zinc plate, and removing any loose deposit from the cells. This constancy of action renders the bat-

tery of great value in the electrotype process, which will be described hereafter. The deposition of metallic copper on the negative plate is the principal inconvenience attending it: this deposit sometimes adheres so firmly as to be difficult of removal, which, however, is only necessary when it interferes mechanically with the working of the battery.

41. GROVE'S BATTERY.—The sulphate of copper batteries, which have been spoken of, have been sometimes called *hydrogen-consuming* batteries, because they provide a chemical agent, oxide of copper, with which the hydrogen produced in the first instance by the decomposition of water, unites to form water again, thus separating copper from its oxide in the metallic form. The only chemical affinity which has to be overcome, in this battery, is that of oxygen for copper. Its force may be stated as equal to the affinity of oxygen for zinc, minus the affinity of oxygen for copper. In Grove's battery, which is now to be described, a further advance is made. The electro-negative metal, as in Smee's, is platinum, which, from the condensed form of the battery, can be used without too great expense. Instead of sulphate of copper, strong nitric acid is employed for the purpose of furnishing oxygen to the hydrogen, with the following advantage: Nitric acid consists of five equivalents of oxygen and one of nitrogen. The fifth equivalent of oxygen is held by a very slight affinity, many chemical agents sufficing to reduce the nitric acid to nitrous acid, which has one equivalent less. The increase of power in Grove's

battery over the sulphate of copper battery, for the same amount of zinc dissolved, is equal to the difference of affinity between oxygen for nitrous acid and oxygen for copper. The force of Grove's battery is, therefore, equal to the affinity of oxygen for zinc, minus the affinity of oxygen for nitrous acid.

42. It should also be remembered that the energy of a galvanic arrangement depends, to some extent, upon the difference in the affinity for oxygen of the metals employed, which, in the case of platinum and zinc, is at a maximum. The zinc plates are amalgamated, and the battery is worked by two fluids; the one within the porous cell and in contact with the platinum being strong nitric acid, and the other, which is in contact with the zinc, being sulphuric acid, diluted with ten or twelve parts of water, as in Smee's, § 25. This battery is the most energetic yet known. A Grove's battery, exposing a surface of zinc equal to 20 square inches, was found, by its magnetizing power, to afford a current of greater quantity than a sulphate of copper battery exposing 210 inches of zinc. The *intensity* of the current in Grove's is also considered three times as great as in the latter, and perhaps four times as great as in Smee's. The same platinum and zinc of a Grove's battery, being used as a Smee's, gave less than a fifth of the current in magnetizing power. Grove's battery is also exceedingly constant.

43. The disadvantages of this battery for common use are, first, the strength of the acids which are employed, making it disagreeable and unsafe to a careless

experimenter; and, secondly, the red fumes of nitrous acid which it gives off abundantly while in action. These are irrespirable and injurious to nice apparatus which may be exposed to them. By placing a metallic covering (protected from the acid fumes) over the battery, and allowing the gases to escape through an orifice stuffed with cotton, wet with a little alcohol, these may be, to some extent, neutralized. It should also be remarked that the *quantity* of electricity produced is, in all these batteries, proportional to the *quantity* of zinc dissolved; and where a small surface of zinc gives a large current, it is acted upon with greater rapidity. The *intensity* of the current produced in these forms depends upon the chemical affinities which are concerned; and on this account there is a gain in the sulphate of copper battery over Smee's, and a still greater in Grove's.

44. The construction of Grove's Battery is shown in Fig. 17. The containing vessel is glass. Within this is a thick cylinder of amalgamated zinc, standing on short legs, and divided by a longitudinal opening on one side, in order to allow the acid to circulate freely. Inside of this is a porous cell of unglazed porcelain, containing the nitric acid and strip of platinum. The platinum is supported by a strip of brass, fixed, by a thumb-screw and an insulating piece of ivory, to the arm proceeding from the zinc cylinder. The amalgamated zinc is not acted upon by the dilute sulphuric acid until the circuit of the battery

Fig. 17.

is completed; but a small portion of the nitric acid filters through the porous cell, by which the zinc would be readily attacked. It is therefore advisable to remove it from the acid when the battery is out of action for any length of time.

45. Fig. 18 represents a series of twelve pairs of Grove's Battery. In this compound battery, each

Fig. 18.

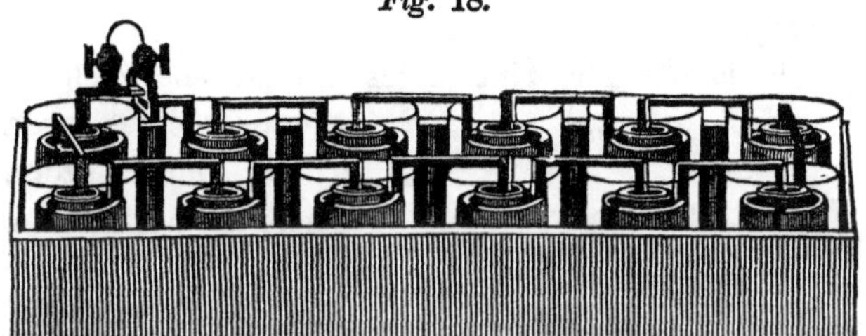

strip of platinum, except the terminal one, is soldered directly to the arm projecting from the adjoining zinc plate. The terminal strip, with its screw-cup, is supported by the first zinc plate, but is insulated from it, as in Fig. 17. Such a series is of very considerable practical importance, as it is now extensively used for purposes requiring a constant galvanic current of great quantity and intensity. It is the form which is in almost all cases resorted to for the magnetic telegraph. For chemical and illustrative experiments requiring a current of great power, this has now, to a considerable extent, superseded the older form of consecutive plates, which will be described immediately. It was stated that Grove's battery may be considered as having three times the intensity of the sulphate of copper battery. A series of twelve pairs, therefore, would be as powerful as thirty-six pairs of the latter; and more so than the same number of pairs in which only sulphuric acid is employed.

46. In Fig. 19, the TROUGH BATTERY, hitherto so much used, is represented. In the form in which it is shown, the plates can be suddenly immersed or withdrawn from the acid at pleasure by means of the windlass. The zinc plates are flat, and inclosed in copper cases, open at top and bottom. There is no division of the acid into separate cells. A represents the trough, B the wooden case containing the plates.

Fig. 19.

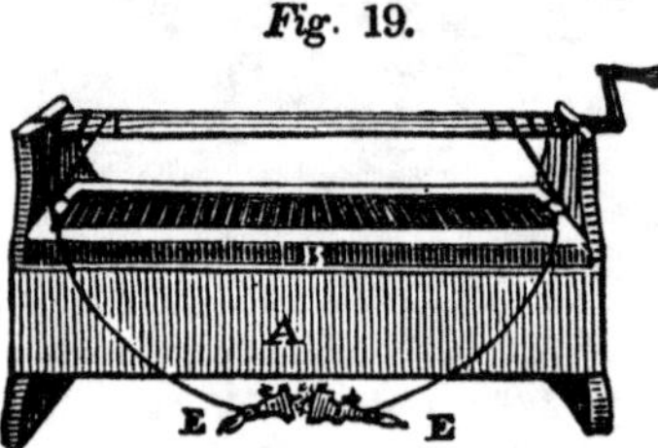

The liquid used is sulphuric acid, diluted with forty or fifty parts of water. If greater power is desired, a little nitric acid may be added. E E are small hand-vices, connected with the poles, for the purpose of holding wires, &c. The battery represented in the cut, consisting of twenty-five pairs of plates, should be able to ignite a considerable length of wire, to decompose acidulated water with rapidity, and to give a brilliant light with charcoal points.

47. Fig. 20 represents a larger battery of a similar construction. There are two distinct series of fifty pairs, each connected with two of the four cups on the table above the battery. In this way the

whole may be used as a single series of one hundred pairs, or as a battery of fifty pairs of double size, by establishing proper connections between these cups. Or only half the battery may be put in action, each having a separate trough to contain the acid. The plates are stationary, and the troughs are raised up to them by means of two racks moved by the crank and handle, H, which lift the platform on which the

Fig. 20.

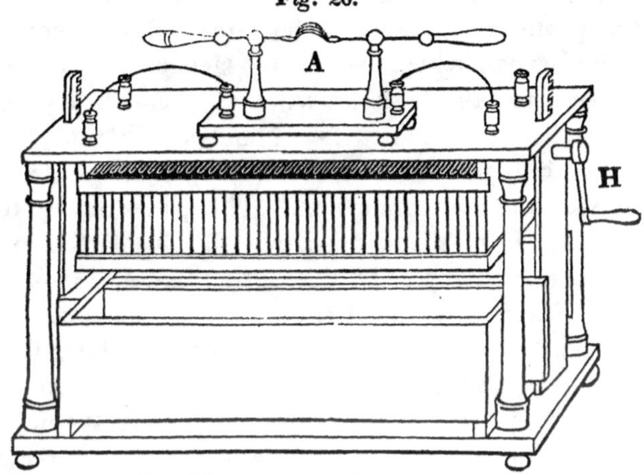

troughs stand. Either trough may be removed from the platform at pleasure, when it is wished to use only half of the battery.

48. In the cut, the arrangement for producing the arch of flame between charcoal points is shown. Two pointed pieces of prepared boxwood charcoal are fixed in the pincers at A, and, the battery being put in action, are brought in contact. The spark passes, and the points become ignited. They may then be

separated to a greater or less distance, in proportion to the power of the battery; and the current will continue to flow through the interval with the production of intense light and heat. This experiment, as well as the deflagration or destructive combustion of metals, is performed with facility by a series of Grove's battery.

49. In the batteries just described, in which the plates are fixed permanently in a frame, the solution of sulphate of copper cannot be employed, on account of the deposit which it forms. Hence diluted acid is used; and the batteries will not maintain a good action for more than a few minutes at a time; in fact, their highest rate of action only continues for a few seconds after immersion. The plates require to be taken out of the acid occasionally during the experiments, and exposed to the air a minute or two.

50. Physiological Effects of Galvanism.— The nervous system of animals is very susceptible to the galvanic influence; so much so, that the sensations and motions, thus excited, are among the most delicate tests of the existence of a galvanic current. It was by observing the contractions in the leg of a frog, when two dissimilar metals, happening to rest upon it, were brought in contact with each other, that Galvani laid the foundation of the science which bears his name. This primitive experiment is shown by the apparatus represented in Fig. 21, where a strip of silver and one of zinc are attached to a little block of wood or ivory.

Fig. 21.

One of the metals is curved over above, so that it may be brought in contact with the other by a very slight pressure of the finger. The lower extremities of the metals approach, and are made to grasp the thigh of a grasshopper recently killed. On making contact between the metals above, the limb will be extended, and on removing the finger, it will again contract. In this case, as in the frog, the exciting fluid of the galvanic pair is the moisture of the skin or the natural fluids of the animal, where the skin is removed.

51. The same apparatus, of larger size, serves to show the contractions produced in the leg of a frog. In this case, a great amount of motion is produced by an almost insensible current; and the only objection to the experiment is, the destruction of life which it requires, and the difficulty of always obtaining the animal. The legs are prepared, as in

Fig. 22.

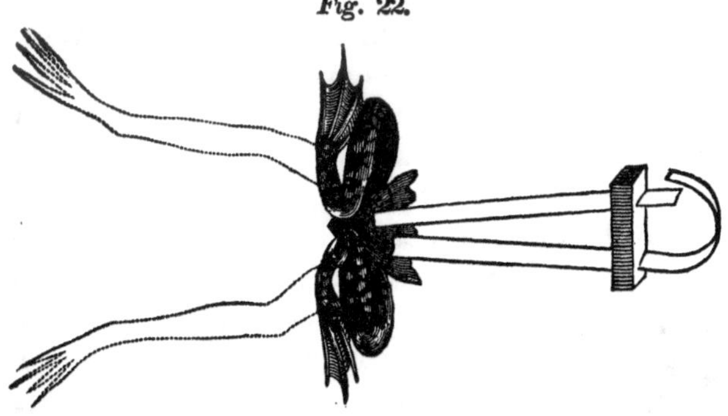

Fig. 22, by removing a small portion of the lumbar vertebræ, with the extremities attached, from a subject recently killed, and stripping off the skin. The legs are shown in the figure in a contracted position, in which they are placed before completing the circuit. The dotted lines exhibit the extended position which they assume under the influence of the current

52. Many animals residing in the water are very sensitive to the galvanic current, owing partly, without doubt, to the large surface they expose to the conducting medium, water. Fig. 23 represents a similar galvanic pair to that described above, introduced into a glass vessel containing a leech; one metal being on each side of the animal. On completing contact, the leech is instantly disturbed, and endeavors to escape from its position in the course of the current. A small fish, in a similar position, also exhibits sensibility; and where a stronger current is passed through the water, as will be mentioned under the head of "Animal Electricity," the creature exposed to it becomes rigid and incapable of motion. A powerful current instantly deprives it of life.

Fig. 23.

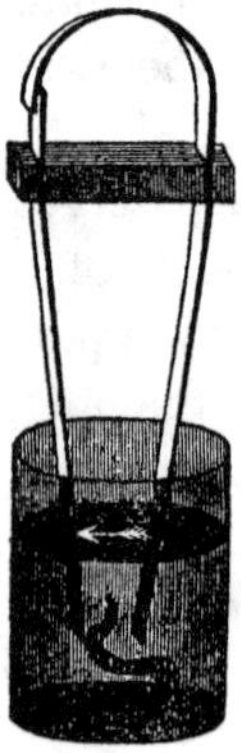

53. If a leech, or fish-worm, be placed on a disc of silver, such as a silver dollar, (Fig. 24,) which rests on a moistened plate of zinc, the animal will show no uneasiness while it remains on one metal, but will instantly withdraw on attempting to pass the limit which separates one plate from the other, with all the appearance of receiving a painful electrical discharge.

Fig. 24.

54. The earliest of all observations in galvanism was undoubtedly the sensation produced by dissimilar metals, when placed above and below the tongue. If the edges of a piece of silver and zinc, so situated, are brought together, a metallic taste and stinging sensation are instantly perceived, the tongue being placed, with reference to the metals, just as the leg of the grasshopper or frog, in Figs. 21 and 22. If one of the metals be placed under the upper lip, instead of on the tongue, the eye will apparently see a flash of light when they are brought together.

55. Conduction of Galvanism. — Metals differ very much in their power of conducting galvanic electricity. The following appears to be the order of conducting power in a few of the most useful metals — silver, copper, brass, iron, platinum ; the first named being the best conductor. For conducting wires or poles, copper is generally used ; for delicate connections, silver. The difficulty of conduction of course increases with the length of a metallic wire, and this becomes an important consideration in the construction of galvanic apparatus. The wires employed with batteries should always be of sufficient size, and of pure and well-annealed metal. It will be found that they lose much in conducting power by being frequently bent.

56. Iron and platinum are used where it is an object to employ the poorest conductors, as in igniting a wire. Galvanic ignition seems to result from a law, that, in proportion to the resistance made by the substance of a conductor to the passage of elec-

tricity, heat is evolved. Thus, when a current of electricity is passed through a metallic wire in greater quantity than it can readily transmit, the wire becomes more or less heated. If its length and thickness be proportioned to the power of the battery, it may readily be melted. A single pair of plates would be the most suitable arrangement for producing this effect, were it not that an increase of intensity enables a greater quantity of electricity to traverse the wire. Hence, for igniting a great length of wire, a battery of a considerable number of pairs is necessary; but a much thicker wire may be ignited by a few pairs of large size. When a very extensive series of small plates is used, the current acquires so high an intensity that its power of producing ignition is diminished, as it becomes capable of traversing a pretty fine wire without obstruction.

57. Either of the batteries of single pairs, which have been described, have sufficient power to ignite a fine wire of iron, or other metal, through which the current is made to pass. This effect is most easily produced in those metals which offer the greatest resistance, not only to the passage of electricity, but also to that of heat; hence a larger wire of platinum may be ignited than of perhaps any other metal, as it is a poor conductor both of electricity and of heat. A steel wire, when intensely heated in this way, burns with beautiful scintillations. The shorter and finer the wire, within certain limits, the greater is the effect produced.

58. Fig. 25 represents an instrument, by means

of which the current may be made to pass through various lengths of wire. The sliding arms may be fixed in a variety of positions, so that their extremities, holding the wire, may be at different distances apart. A fine wire, if not too long, thus interposed in the course of a powerful current, is immediately fused or deflagrated. This instrument has been used to determine the conducting power of metals, by passing the current of a constant battery through an instrument capable of measuring its magnetizing power, and then also opening a collateral circuit to the same current through a certain length of fine wire, first of one metal, then of another. The instrument employed to measure the magnetizing power was the *Magnetometer*, which will be described hereafter. In this instance, the current is divided a portion of it, equivalent to the conducting power of the metal under experiment, passes through the fine wire; and the measuring instrument, through which the remainder of the current passes, indicates a corresponding diminution. When the whole current was made to pass through the instrument, the magnetizing power in one experiment was 24,400 grs.; that is, the electro-magnet employed attracted

Fig. 25.

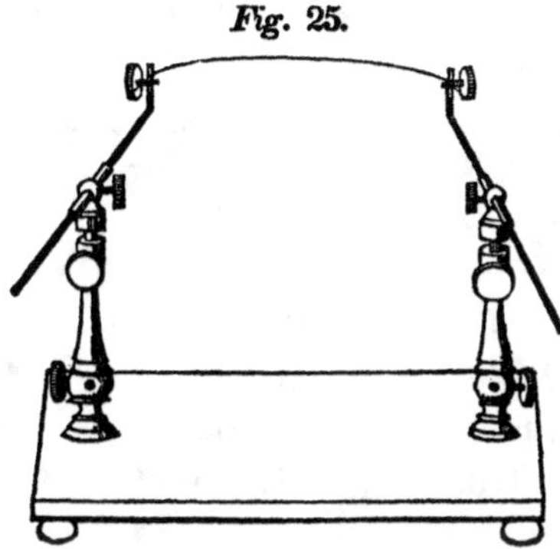

its armature with a force equivalent to this weight. The following table will show the reduction of this amount, when fine wires of various metals were employed in the way above described.

	Magnetizing Power.	Reduction.
Silver coin, .	7,350 grs.	17,050 grs.
Copper, . .	8,180 " .	16,220 "
Pure silver, .	8,800 " .	15,600 "
Brass, . . .	12,046 " .	12,354 "
Gold coin, .	12,820 " .	11,580 "
Iron, . . .	16,600 " .	7,800 "
Platinum, . .	17,010 " .	7,390 "

The conducting power of the metals is, of course, in proportion to their reduction of the magnetizing power of the collateral current. The alloy of silver with a little copper used in coinage, is the best conductor of the metals tried, and platinum the poorest.

59. Powder Cup. In fig. 26 is represented a little instrument designed to show the heating power

Fig. 26.

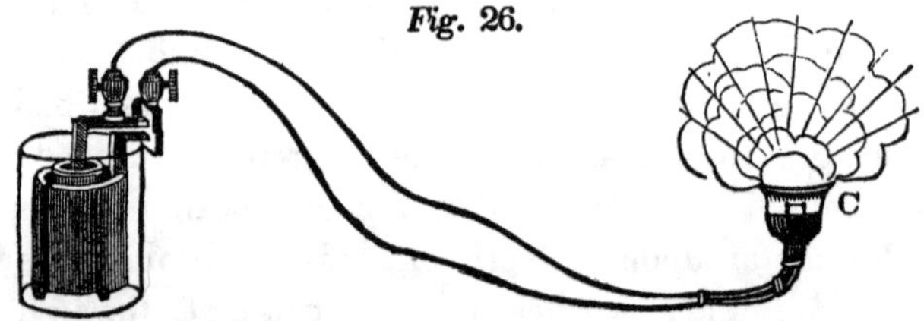

of the battery current. Two copper wires, wound with cotton thread, except at their ends, are joined by a short piece of fine platinum wire. These wires pass through the bottom of a small glass cup, C, so that the platinum wire lies free in its cavity. On putting a little gunpowder into the cup, C, and then connecting the copper wires with the poles of the battery, the platinum will become heated, in consequence of the flow of the current through it, so as to inflame the powder. The recent discovery of gun-cotton has a very useful application to this instrument, and to the galvanic battery generally, as a means of igniting explosive compounds. The temperature at which gun-cotton explodes being only about 360° F., requires a comparatively slight elevation of temperature in the wire. For blasting or submarine explosions, this will be an important advantage.

60. THE VOLTAIC GAS PISTOL, represented in Fig. 27, is constructed on the same principle as the last-described instrument. The wires pass up through a brass piece which screws into the barrel, they being completely insulated from each other. A sectional view of this part is annexed. The fine platinum wire stretches across between the two. The introduction of a proper quantity of hydrogen gas may be effected in the following manner: Connect with a self-regulating reservoir of hydrogen a leaden or other tube, so bent as to deliver the gas under the surface of water in a jar. The pistol being uncorked, and the brass

Fig. 27.

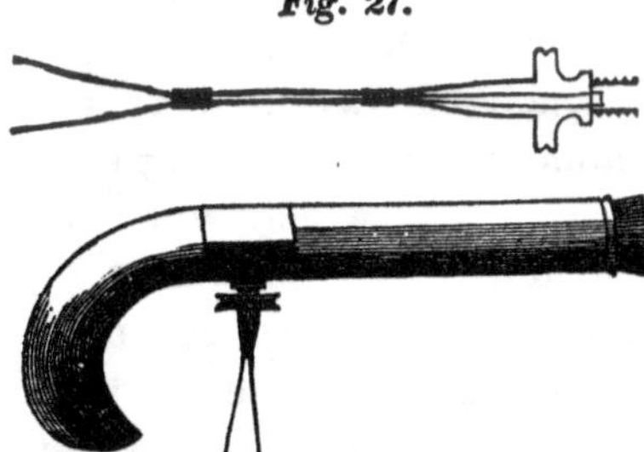

piece unscrewed, immerse the muzzle in the jar to such a depth that the water may fill one quarter of the barrel. Then replace the brass piece, and, bringing the muzzle over the end of the tube, open the stop-cock of the reservoir. When the escape of bubbles shows the pistol to be full of gas, withdraw it, and insert the cork. In this way it will contain one volume of hydrogen to three of air, which is the best proportion. If too much hydrogen is introduced, no explosion will occur. It is not, however, necessary to be very particular; and it will answer the purpose almost equally well, without going through the troublesome process above, if the pistol is held for a few moments over a jet of the gas. The explosion is louder and more certain to occur, if it is filled with a mixture of oxygen and hydrogen, in the proportion of one volume of the former to two of the latter.

61. The pistol being corked, connect one of the wires with the zinc pole of the battery, and the other with the copper pole. The circuit will now be completed through the platinum wire; this will instantly be ignited, setting fire to the gas, which will expel the cork with a loud report. The removal of the brass piece and wires allows the mixed gases to be fired by the application of flame when desired.

62. Galvano-Decomposition.—It has been stated, that the liquid between the plates of the battery undergoes decomposition as a condition of the passage of the galvanic current. It was also stated that this took place not only in the battery

but whenever the current produced by it was made to pass through a liquid consisting of more than one element. In this case, it is found that a certain class of substances always appear at one pole of the battery, and a certain class always at the other pole. Thus, when solutions of metallic or other salts are decomposed, the metal, or base, always appears at the zinc, or negative pole; while the electro-negative element, previously in combination with the metal or base, as chlorine, oxygen, cyanogen, appears at the other, or positive pole. The decomposition is exactly proportional to the quantity of electricity which passes; and instruments have been contrived to measure the current, by measuring the amount of matter decomposed by it. It is also found that, for a given amount of electricity, a definite proportion of each element is evolved; and these proportions are the same as the chemical equivalents of those elements. Substances differ very much in the facility with which they are decomposed. Some form the most delicate tests we possess of the passage of an electric current. Others require the most powerful galvanic apparatus to separate them into their elements.

63. If the poles of a battery of two or three pairs are made to terminate in strips of platinum, and to dip into water, a slender stream of gas will be seen to arise from each wire, hydrogen being evolved from one, oxygen from the other. If the wires are of iron or copper, the oxygen will unite with the one connected with the positive pole to form oxide of iron or of copper, and hydrogen alone will be given off.

Platinum wires or strips are not attacked by the oxygen, and are therefore best for the decomposing surfaces. The decomposition of water is greatly facilitated by dissolving in it some salt, as, for instance, Glauber's salt, or, what is still more effectual, by the addition of one part of sulphuric acid to ten or fifteen of the water. These substances increase its conducting power. If, however, the power of conduction be increased beyond a certain limit, the evolution of gas will be lessened.

64. Fig. 28 represents a Decomposing Cell, mounted on a stand. Two platinum wires, connected with the cups A and B on the stand, pass up into the cell, which is of glass. A glass tube, G, may be inverted over these wires, to collect any gas which is evolved; it passes through a cork, fitting the mouth of the cell with sufficient tightness to allow the tube to be filled with the liquid by merely inverting the instrument. The cell, being partly filled with acidulated water, and the tube, wholly full, connect the cups, A and B, with those of the battery. Immediately bubbles of gas will be seen to escape from each wire, and to rise into the tube, displacing the liquid from it. When the tube is full, it may be re-

Fig. 28.

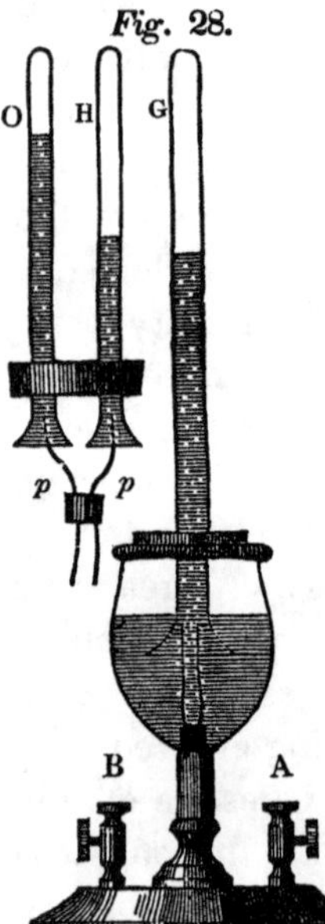

moved and the mixed gases exploded by holding its mouth to a flame.

65. By having two glass tubes, O and H, passed through a cork, so that one of them may be inverted over each wire, as shown in the cut, where *p p* are the platinum wires, the gases may be obtained separately; oxygen only being collected in the tube placed over the positive wire, and hydrogen alone in the other. The volume of the latter gas is twice that of the former, as indicated in the figure by the relative height of the liquid in the tubes O and H, occupied respectively by the oxygen and hydrogen. On removing the tubes when full, the hydrogen will burn if a flame be applied, and the oxygen will increase the brilliancy of the combustion of any ignited body put into it. When these tubes are graduated, they constitute the instrument called *voltascope*, or voltameter. The amount of gas liberated is exactly proportioned to the electricity which passes, and affords the most exact measure of the amount of a current when it continues for any length of time.

66. A solution of iodide of potassium is perhaps more easily decomposed than any other salt. In this case, instead of the elements of water, the components of the salt itself appear at the poles; iodine at one side, and potash, by a reaction between potassium and water, at the other. When iodine and starch are brought together, an intensely blue precipitate results. Melloni availed himself of this fact, to determine the heat of insects, in some of the most delicate experiments ever made, by means of his thermo-

electric pile. A little starch being added to the solution of iodide of potassium, the passage of the slightest electrical current is made apparent by the blue precipitate at the positive pole. The direction, as well as the existence, of a current is thus shown. The decomposing cell (Fig. 28) will answer for this experiment; but a more delicate form is shown in Fig. 29, where a glass tube is supported on a stand, two platinum wires, each connected with a small screw-cup, entering on opposite sides. These wires are insulated, except at the extremities, which are flattened into little discs, and are parallel to each other. When a current of almost inappreciable quantity passes, as that from a galvanic pair composed of a zinc and a silver wire, whose extremities are dipped into distilled water, or from two or three turns of a small electrical machine, one of these discs will be seen, when compared in the light with the other, to become of a purple hue, the effect constantly increasing as the current is continued. The disc recovers its brightness by stirring the solution near it, or on being touched with a little splinter of wood.

Fig. 29.

67. Most of the salts having alkaline bases are decomposed into an acid and an alkali, which appear at the opposite poles. Some of the phenomena occasioned in this way are best observed by placing a porous diaphragm across the decomposing cell, so as to form separate apartments for the poles. Fig.

30 represents a tumbler divided vertically; with the edges ground together, so as to form a tight joint,

Fig. 30.

when they include between them a diaphragm of unsized paper. They are held together by rings of India rubber. A little screw-cup is clamped to the side of each of these apartments, from which platinum poles descend into the vessel.

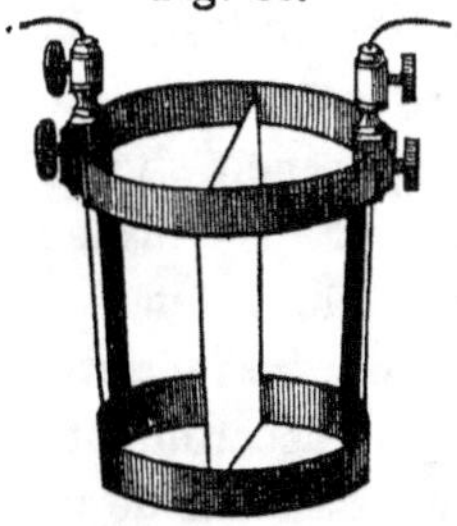

68. Let this apparatus be filled with a solution of Glauber's salt, (sulphate of soda,) to which has been added enough of the infusion of red cabbage to give it a blue color. On the passage of the current from a single pair, the blue color will soon be changed to red in the cell containing the positive pole, and to green in the other, sulphuric acid and soda being respectively evolved. By reversing the current, the original color will be first restored in each cell, and then the opposite change will occur. If the solution is colored blue by the infusion or tincture of litmus, it will become red in the cell in which the acid is developed, but will suffer no change in the other. When the yellow infusion of turmeric is used, it is turned brown by the alkali evolved in the negative cell, but is not affected by the acid in the other. Similar phenomena present themselves with a large number of salts, but in some cases different effects are produced.

69. If a solution of muriate of ammonia, colored by some vegetable infusion, be employed, chlorine

gas will be given off from the positive pole. This may be recognized by its peculiar odor, and by the bleaching effect it produces upon the liquid in the positive cell, which quickly becomes colorless. In this case, ammonia and hydrogen are set free in the negative cell, and muriatic acid and oxygen should have been liberated in the other. The chlorine appears therefore to be a secondary product, set free by the combination of the hydrogen of the muriatic acid with oxygen, to form water. Or, according to the view which regards sal-ammoniac as a chloride of ammonium, chlorine is directly liberated at the positive pole; and at the negative, ammonium, which, not being able to exist in an isolated state, is instantly resolved into its components, ammonia and hydrogen.

70. In most cases, it is desirable to have the distance between the poles in the decomposing cell as small as possible, as the current passes more easily and the action is more rapid. Where, however, different colors are to be exhibited, the instrument represented in Fig. 31 may be used with advantage. This is a U-shaped tube, into which the poles enter at the top. It may either form a continuous cell, or it may be divided by thrusting a piece of loosely crumpled, unsized paper, or of cotton cloth, into the bend; or the

Fig. 31.

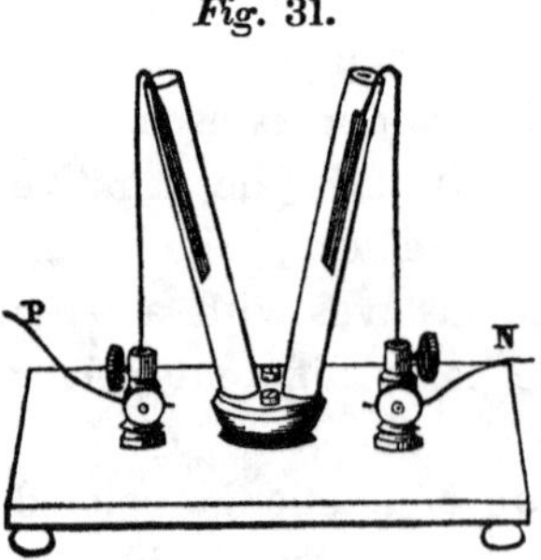

tube may be in two parts, cemented together into a brass support or stand, with a porous diaphragm between. When the experiment with sulphate of soda, § 68, is tried with this instrument, three separate zones of color, blending into each other, are perceived. At the positive pole, it is red; at the negative, green; and in the region between, the color remains blue. As the experiment proceeds, the red and green solutions encroach on the blue, until a small space only remains between them, if the tube is continuous, where the acid and alkaline portions blend together.

71. Electro-Metallurgy. — Solutions of metallic salts undergo decomposition in the same way as the alkaline salts. The result is the deposition of the metal itself at the negative pole. Even with the alkalies, a battery of great power, in place of the alkaline oxide usually evolved, separates the metal which is its ultimate base. The decomposition in alkaline and metallic salts takes place, probably, in the base of the salt, the metal being separated at one side, and the oxygen, or electro-negative element, at the other. In some cases, as with the alkalies, the elements thus eliminated decompose water, and an acid and alkali finally appear at the poles. The deposition of metal has been spoken of in the sulphate of copper battery. The observation of this important fact, in its connection with the mechanic arts, was made at about the same time, in 1837 and 1838, by Jacobi, of St. Petersburg, and Spencer, of Liverpool. The science of *The Electrotype*, or *Electro-metallurgy*, thus had its origin.

72. The great object in depositing metals is to throw them down in a malleable, coherent state — the *reguline* condition. If different metallic solutions should be put in one of the decomposing cells already described, it will be found that copper will probably be deposited on one of the poles with its proper color, while silver and gold will be thrown down as a black powder. A great difference, therefore, exists with the different metals. Some practical directions will be given for the precipitation of those most important.

73. Metals tend to be thrown down in three different states — as a black deposit, as reguline metal, and in a crystalline condition. These forms all depend upon the fact, that in the decomposing cell there are two bodies present, water and the metallic salt, either or both of which may be decomposed by the galvanic current. Now, when water is decomposed, hydrogen is evolved at the negative pole; and when this takes place at the same time that the metal is evolved at the positive pole, the black deposit occurs. Smee suggests that the hydrogen and metal are both thrown down together, and, the mixture being incompatible, the metal appears in an infinitely divided state. As an illustration of this principle, if a metallic solution is acidulated, the water will be more easily decomposed as the acid increases its conducting power. There will, then, be a greater tendency to the evolution of hydrogen. Where a metallic salt is difficult of decomposition, acid is sometimes, therefore, added; where a black deposit is

reared, it is avoided. In Smee's acid battery, we have seen hydrogen evolved at the negative plate; in the sulphate of copper battery, we have seen copper deposited on the negative plate, but in a hard and crystalline condition. Now, unite these by adding sulphuric acid to the sulphate of copper solution, until you arrive at the point when, copper being still deposited, hydrogen is just about to appear, and the metal will be precipitated in a soft and malleable form. The crystalline state results from the deposition of the metal in obedience to its own molecular attraction. In the reguline state, this is modified by the chemical agency of the elements of water.

74. The rule prescribed by Smee is in all cases to regulate the strength of the battery current according to the strength of the solution. The following formula for controlling the evolution of hydrogen, and thereby regulating the quality of the metal, is condensed from his work on electro-metallurgy. The tendency to the evolution of hydrogen, and to a black deposit of metal, is increased by any of the following means: By increasing the intensity or quantity of the battery; by increasing the size of the positive pole; by decreasing the size of the negative or depositing pole; by approximating the poles; by increasing the heat of the solution; by weakening the solution; by making it acid. The tendency to evolution of hydrogen may be diminished, and that to the crystalline form of deposit increased, by any of the following means: By diminishing the intensity or quantity of the battery; by increasing the

negative pole; by decreasing the positive pole; by separating the poles; by saturating the solution; by making it neutral; by diminishing its heat. It will frequently be found that one or more of these means may be employed in a particular operation, while it will be impossible to make an advantageous application of others.

75. The metal of the electrotype preserves the precise form of the negative pole or mould upon which it is deposited. A finger-mark, or slight tarnish on the mould, is faithfully copied by the surface of the metal thrown down in contact with it. The chief applications of this art at present are, the reproduction or manufacture in copper of medals, dies, engraved plates, plaster casts, ornamental raised work, and generally of all designs in relief or intaglio. Daguerreotype pictures are also reproduced on a copper surface by the same means. Another important application is the coating of the more oxidizable metals with films of the noble metals, as in gilding and silvering. If a sheet of copper should be thrown down on an object which it was desired to copy, and then be separated from it, a reverse impression of the original would be obtained. To obtain a fac-simile of the original, it would be necessary to use this reverse as a mould, and proceed to precipitate metal upon it, as in the first instance. This is accordingly one of the processes used to obtain reverses or moulds of objects to be copied.

76. For this purpose, a medal or engraved plate is placed itself in the solution and copper deposited

upon it. The negative wire of the battery should be connected with the rim of the medal, and in case of an engraved plate, it may be soldered to the corners. The deposit is apt to adhere very firmly, sometimes so much so that its removal is impossible. This may be avoided by slightly greasing or oiling the mould, and then brushing it over with a little dry copper bronze.

77. The mould thus obtained may have a wire soldered to it, and be placed in the solution like the original one. In most cases, it will be considered safer to take a mould of a valuable medal or plate in soft wax, or by some of the other processes to be described.

78. An engraving printed from an electrotype plate obtained by this process, is given, as a specimen of the art, in the frontispiece to this Manual, in company with one from the original copperplate. No difference can be detected between the impressions, except that arising from the greater or less quantity of ink left in the work, as occurs in different engravings printed from a single plate. This is an important application of the art, as in cases where a large number of impressions are required, or where it is desirable to print from a plate in different places, two or more plates have been obliged to be engraved, while now it is only necessary to engrave one, which will not be injured in the slightest degree by taking copies from it, in one or other of the modes. Steel plates must not be placed in the copper solution without a modification of the process which is given

in § 96. It was found that the electrotype plate used in the first edition, wore better and worked better than the original. This is due to the great purity and equality of the deposited copper and the hardness which can be obtained by regulating the current so that there shall be the slightest tendency to the crystalline deposit.

79. One of the early processes for obtaining moulds is that called abroad the *clichée* process, which consists in stamping the object to be copied on fusible alloy just at the moment of hardening. This alloy consists of three parts of tin, five of lead, and eight of bismuth, and melts at about the temperature of boiling water. Rose recommends an alloy of one part tin, one part lead, and two of bismuth, which fuses at about 200° F. When the alloy is melted, it is skimmed with a card, and the medal or plate, attached by wax or otherwise to a handle, is suddenly and forcibly pressed upon it. When of small size, a mould may be made, by a few trials, presenting an accurate reverse of the original. The frontispiece to the first edition was copied in this way. A clean copper wire, to be used as the negative pole of the battery, is heated at the end with a little rosin upon it, sufficiently to melt and solder itself to the edge of the fusible mould, which is then ready for the solution. All parts on which the metal is not to be deposited are protected by a coating of varnish or wax.

80. The moulds heretofore described have been formed of galvanic conductors. Means have also

been found of using moulds of the important class of non-conductors, as wax, sealing-wax, and plaster of Paris. A mould is easily made of a soft composition wax by bringing it to an even surface, and then pressing the object to be copied upon it with sufficient force. The surface of the original plate should be slightly oiled and brushed with blacklead, to prevent its adhesion. In this way, very perfect moulds, even of large objects, are obtained, and it will be found the simplest and best mode of operating in most cases. Medals, copperplates, wood cuts, pages of type, may be thus copied. The page in which these remarks occur, has been electrotyped by this process, and may be compared with the others, which are stereotyped. Parts of wood cuts in this volume which it was desirable to insert in more than one place, have been multiplied in the same way. Melted wax may also be poured on surfaces which are to be copied; but this is more troublesome, and less successful than the mode already described.

81. It is necessary to render the surface of the wax mould a conductor of electricity. This is done by giving it a coating of good blacklead, which should be rubbed over its face with a soft brush, until it acquires a shining black appearance; a very thin film is sufficient. A copper wire is then to be bent, so as to conform nearly to the edge of the mould in its whole circumference, and then to be heated in a lamp, and pressed upon the wax, when it will become imbedded. Communication between the wire and the face of the mould is to be insured

by rubbing a little blacklead on the parts around the wire.

82. The mould, when thus prepared, may be put into the solution, care being taken to remove air-bubbles. The deposit commences upon the wire, and extends gradually over the blackleaded portions. The mould may be employed again, if a new coating of blacklead is given to it. The metallic moulds can also be used any number of times, if uninjured.

83. Moulds are obtained from plaster medallions by pressing them upon a matrix of soft wax, after their pores have been saturated with water. The earlier process consisted in soaking them with hot water, and then pouring on wax and allowing them to cool before separating it. Reverses of plaster casts are obtained by soaking them in hot stearine, tallow, or wax, as long as any bubbles of air escape, and then removing all excess of these substances from the surface, and blackleading as before. There is, however, a serious objection to placing plaster casts in a solution, from the fact, that, if the smallest quantity of sulphate of lime is dissolved, it affects very injuriously the deposited metal.

84. Where it is not necessary to get a precisely accurate copy of a surface, it can be made a good conductor of electricity by giving it a coat of copper-bronze varnish. Copper spreads at once over this when placed in the solution. Fruit may thus be easily covered with a metallic crust, preserving its form and characteristic appearance.

85 Impressions of seals may be copied by a very

simple process. They are to be covered with a thin film of blacklead, rubbed on with a soft brush. If this does not adhere readily, the brush may be very slightly moistened with alcohol, care being taken not to roughen the surface of the seal. A wire is then melted into the sealing-wax, and the seal placed in the solution. The operation is similar in all respects to that required for the wax moulds. The original seal or stamp may be copied by taking an impression in soft wax.

86. Where the mould is of copper, a single cell has been sometimes made to do the work of both battery and depositing apparatus. The mould is placed on one side of a porous partition in the copper solution, and a wire connects it with a piece of zinc on the other side, in a solution of sulphate of soda. The action in this case cannot be nearly so well controlled as where a separate battery is used, and the deposited metal is not as good.

87. To copy small objects, a single Smee's battery is sufficient. Where a large surface is to be deposited on, it requires the superior energy of the sustaining sulphate of copper battery. The leather cell here has an important application, as it can be adapted in shape to masses of spelter or commercial zinc, which is the most economical form in which that metal can be employed in the large way.

88. With the copper plate of the battery is connected a piece of copper which is to be immersed with the mould in an acidulated solution of blue vitriol, contained in a glass, or well-glazed earthen-

ware vessel, or in a wooden trough. The piece of copper and the mould should both be connected with the battery, and the copper placed in the solution, before the mould is introduced; in this way all local action is prevented, and the deposition of copper commences immediately. Any air-bubbles which adhere to the mould must be dispersed. The solution is prepared by adding 2 oz. of sulphate of copper and 1 fluid oz. of sulphuric acid to every 15 fluid oz. of water. As the copper is deposited on the mould, an equal quantity is dissolved off from the immersed plate, so that the original strength of the solution is maintained except for the loss of water by evaporation.

89. Fig. 32 represents a Smee's battery, B, of the mercurial, or "odds and ends" form, a new and improved variety, connected with a depositing apparatus. A little mercury is placed in the bottom of the glass vessel forming the battery. A platinum plate is seen suspended in the centre, beneath one of the screw-cups. One or more zinc plates rest against the side of the vessel, and are in contact with the mercury below. A curved wire is seen descending through the liquid, insulated by a glass ube, to the mercury, which it connects, in

Fig. 32.

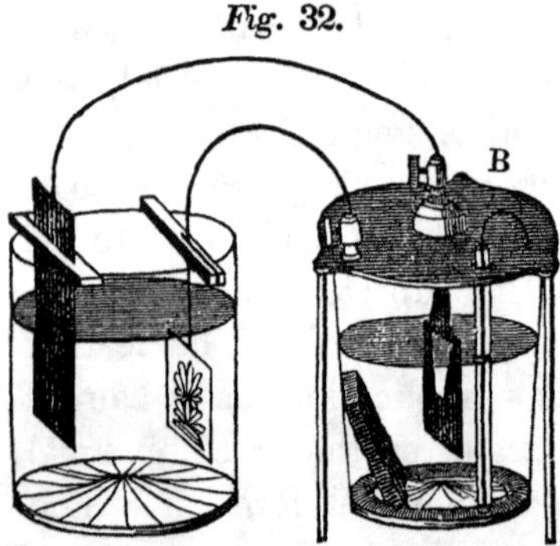

common with the zinc, with the plate on the top and the other screw-cup. This battery will operate if simple scraps of zinc are placed in the mercury. The mould in the depositing cell will be seen to be connected with the zinc or negative pole of the battery.

90. During the solution of the positive plate, a considerable quantity of black matter is left, mostly carbon, which would injure the copy if allowed to fall on the mould. It is, therefore, best to place both in a vertical position, the face of the mould being opposite the piece of copper. The solution may be stirred occasionally, to keep its upper and lower parts of equal strength.

91. When the process is going on well, the deposited metal will be of a very light copper color. The rapidity of the deposition depends greatly upon the temperature; the process proceeds much faster in warm weather than in cold, and still more so if the solution be kept hot. A thickness of one tenth of an inch may require from three days to a week for its formation, when artificial heat is not used. When a sufficient thickness has been attained, the copy may generally be removed from the mould without difficulty, care being taken to cut away any copper which embraces the mould at the edges.

92. Every ounce of copper deposited requires the solution of somewhat more than an ounce of zinc from the zinc plate of the battery. Five or six electrotypes may be made at once, without increasing this expense, by arranging in succession several vessels, each containing a mould and a copper plate con

nected by a wire with the mould in the next one The plates of copper and the moulds should all be nearly of the same size, and the solution should contain less blue vitriol and more sulphuric acid than directed in § 88, particularly if the series extend beyond two or three. When the moulds are small, glass tumblers form the most convenient vessels. Fig. 33 illustrates this arrangement, in which one battery is seen connected with three depositing cells.

Fig. 33.

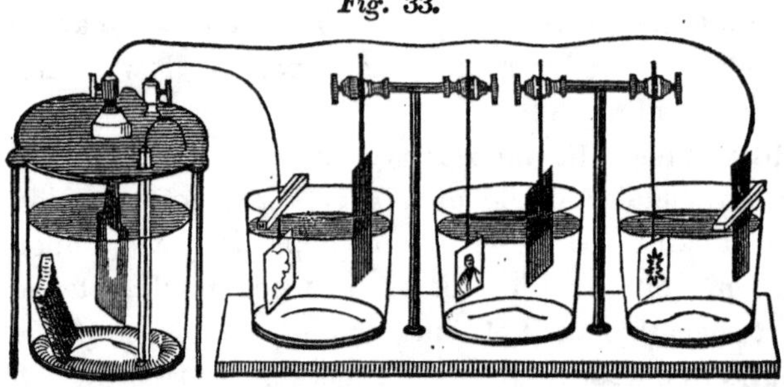

The moulds will be observed to occupy the same position in relation to the current in each of the cells. The copper plates opposed to the moulds are dissolved in proportion to the amount of copper deposited, so that the solutions retain their strength by the recomposition of sulphate of copper, exactly equal to the decomposition at the surface of the moulds. The increase of the number of cells requires, therefore, no increase of electrical energy, except for the greater distance through the solutions which the galvanic current has to pass. In this way several ounces of copper are obtained, with but a slight increase in the quantity of acid or blue vitriol required for working the battery, and a little more corrosion of the zinc plate.

93. The surface of the electrotype copy is usually of a bright copper color; but sometimes it presents a dull surface. If it is discolored, it may be cleansed by immersion for a few moments in dilute nitric acid, and then washed with water. It may be bronzed by brushing it over with blacklead immediately upon its removal from the solution, and having heated it moderately over a clear fire, by rubbing it smartly with a brush, the slightest moisture being used at the same time, in order to remove the black lead. It may also be gilt or silvered.

94. If the copper plate which is opposed to the mould in the solution is coated with wax, in which lines are drawn reaching the metal, it will be etched by the acid, and may afterwards be printed from like a plate etched in the usual way by nitric acid. There is the peculiar advantage in this process, that the action is perfectly regular over the whole plate, or that a gradation of action may be produced by approaching the negative pole to the parts where the shades require to be bitten in deepest. There are also no nitrous fumes evolved, as with nitric acid. A plate of electrotype copper is preferable to any other for etching, as its purity insures an equal result. The negative pole should be small, and at some distance from the plate to be etched, when it is intended that it shall act uniformly.

95. A process of surface-printing, called Glycog-

raphy has recently been devised, by a reverse operation to that of galvanic etching. A design is drawn through an etching ground upon a polished plate, as before; the whole is then blackleaded, and a plate of copper thrown down, in which the lines of the drawing are of course raised like those of a wood cut. This plate is printed from, like an engraving on wood, in the same page with common type; which cannot be done with a copper plate or steel plate engraving.

96. It is sometimes desirable to precipitate copper on iron or steel, to prepare it for receiving the noble metals, or for acting as a mould in the common sulphate of copper solution. The solution used for this purpose is the cyanide of copper, which is not decomposed by iron, and which is prepared by dissolving oxide of copper, or sulphate of copper, in a solution of cyanide of potassium. This salt is only useful for throwing down a film of copper upon iron or other oxidizable metals. The surface of the iron must be perfectly clean.

97. Copper is the only metal which can be used with advantage for the purposes which have been described. The precipitation of the noble metals, silver, gold, and platinum, is of great importance for purposes of ornament and protection to other metals. Their salts give water a high conducting power, and are themselves easy of decomposition. There is, therefore, a great tendency to the evolution of hydrogen, and this evolution is peculiarly liable to reduce the metal as a black powder. It is as yet

almost impracticable to reduce platinum in other than the divided state, which has, however, an important application in the coating of platinum plates for batteries, where it prevents the adhesion of hydrogen. The solution used for this purpose is the chloride obtained by dissolving platinum in nitro-muriatic acid, and partially evaporating to expel any great excess of acid. The solution may be subsequently diluted to any required amount.

98. The battery used for gilding and silvering consists of from three to six or eight pairs, according to the extent of surface to be covered, and the resistance offered by the solution. As the solutions grow older, they require less power to decompose them; but there is then a greater tendency to the black deposit. The positive pole usually employed is a platinum wire, or pointed strip of platinum foil This is not dissolved by the solution, and consequently the electro-negative element of the salt has

Fig. 34.

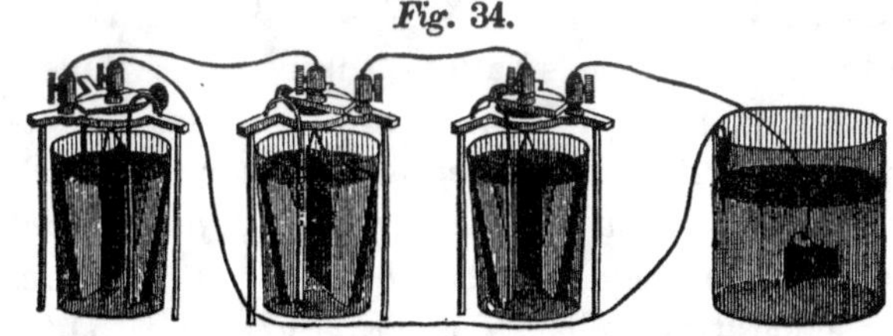

to be evolved, instead of entering into a new combination. It is on this account that several pairs have no more efficiency in decomposing the solution than one pair where the positive pole is dis solved. The object of the present arrangement is to bring the whole action more exclusively under the control of the battery, and to avoid local action. For ordinary purposes, a Smee's battery, weakly charged, is preferable. Fig. 34 represents an odds and ends battery of this kind, connected with the silvering apparatus, differing only from that described in § 89 by being uncovered. It will be observed that the positive pole, in the decomposing cell, is placed at a distance from the object to be silvered.

99. Another form of Smee's battery, connected with a depositing apparatus for silvering daguerreotype plates, is exhibited in Fig. 35. This is the

Fig. 35.

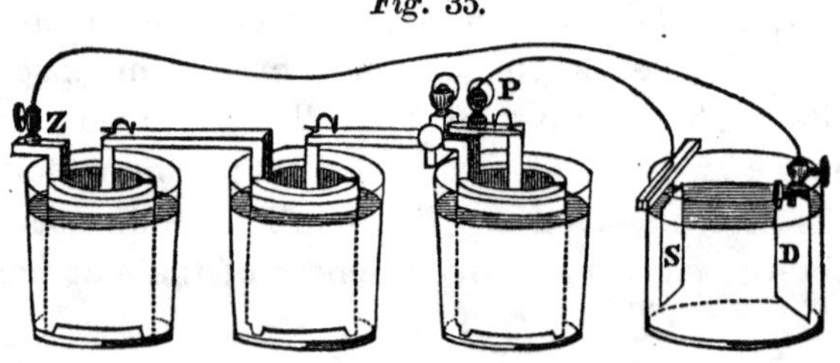

same form with Grove's battery, the porous cylinders only being removed. The platinum plates are attached to little arches of wire, which enter small holes in the zinc, into which they may be wedged or from which they may be removed at pleasure. In the depositing cell, a daguerreotype plate, D, is seen attached by its edge to a screw-cup and clamp prepared for that purpose, and thence connected with the terminal zinc cylinder of the series, Z. Opposed to the daguerreotype plate, is a plate of silver, S, connected with the platinum pole, P, of the battery, and which is dissolved by the solution in proportion as silver is deposited on the daguerreotype plate. A very pure coating of silver is thus given to a plate from which the best pictures are obtained.

This fig. shows the porous cell belonging to the battery when used as a Grove's. The use of this cell, and of nitric acid within it, constitutes the difference between these forms.

100. Fig. 36 represents a series of the sulphate of copper battery applied to gilding and silvering.

Fig. 36.

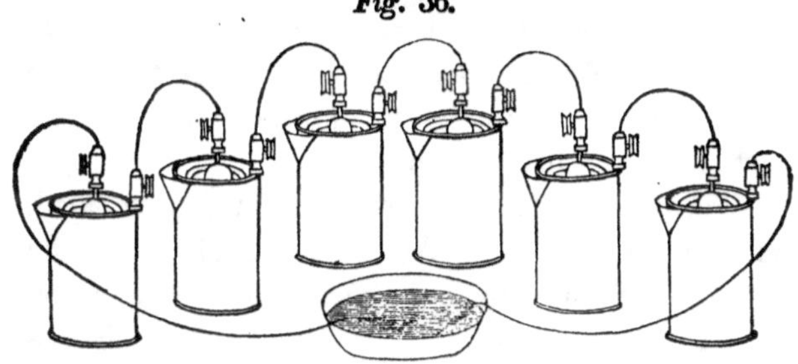

This may be used where a very large surface is to be deposited upon. The figure is intended to illustrate chiefly the deposition of alloys by the galvanic current. If two metals are contained in a solution, the general law is, that the one most easily reduced by the electrical process will be deposited first, and in a state almost absolutely pure. If the energy of the current, however, is very much increased, all the metals present will go down in variable proportion. Thus, if there is a little silver in the gold electrotype solution, a feeble current will throw down the silver first; if there is copper present, and no silver, a feeble current will throw down a pure yellow deposit of gold, while a stronger one will throw down a reddish metal resembling the gold of jewellers and of the mint.

101. The salt of silver, used for precipitation, is the ferrocyanide. A silver dollar is dissolved in nitric acid, and the silver precipitated in the state of chloride, by muriatic acid or common salt. The precipitate is washed, and then added to a sufficient quantity of a hot saturated solution of ferrocyanide of potassium (yellow prussiate of potash), to dissolve it. Sufficient water is then added to make two quarts. The solution may be poured off from the sediment which remains, or it may be used at once. It must be remembered that the salt employed contains cyanogen, the active principle of prussic acid. Prussic acid itself and oxygen are evolved at the positive pole during the decomposition. A solution of cyanide of silver may also be used, which is obtained by dissolving the precipitate of chloride of silver, mentioned above, in cyanide of potassium, instead of the ferrocyanide. In this case, the solution of cyanide of potassium need not be so strong, or be raised above the common temperature. Only sufficient of the cyanide of potassium need be added to take up all the chloride of silver.

102. The solution of gold employed is the cyanide. This is made by dissolving a five dollar piece in nitromuriatic acid (a mixture of one measure of nitric acid with four of muriatic), and evaporating the salt very carefully, in order to drive off the excess of

acid. This is then redissolved with a sufficiency of cyanide of potassium to give a clear solution, it requiring somewhat less than an ounce of the cyanide for this purpose. Enough water is added to make two quarts. As a general rule, it requires more battery power for gold than for silver. The addition of a small quantity of cyanide of potassium to a solution of gold diminishes the tendency to black deposit where this is exhibited, and where the fault is referable to the solution.

103. It is necessary that the surface on which a metal is deposited should be perfectly clean, and free from a film of air, that it may adhere; and the metal should be thrown down with sufficient energy to prevent any local action with the mould. The article to be silvered or gilded may in some cases be cleaned by a solution of potash; but the deposit is more certain to adhere if the surface is rubbed with a little rottenstone when first placed in the metallic solution, and connected with the battery. Whenever, during the process, the deposit becomes discolored or rough, the negative plate should be taken out, and brushed with a little whiting and soap and water. The thickness of the coating of gold and silver is proportional to the time occupied by the deposition and the amount of electricity which passes. For the same amount of electricity, 32 grains of copper, 108 grains of silver, and 99 grains of gold, are deposited, these being the chemical equivalents of those metals.

104. If a small object, such as a dial of a watch,

be placed in a solution of cyanide of silver or gold in a capsule, the metals will be thrown down upon it in a very satisfactory manner, if a small slip of zinc be placed in contact with it, as represented by Z in Fig. 37, and the whole be heated by a spirit-lamp. It is necessary that there should be rather more cyanide of potassium in the solution than in that for the ordinary process. The dial of the watch itself becomes here one plate of the battery, and the slip of zinc the other. The zinc is dissolved by the solution; but by the law regulating the deposit of alloys, only gold is thrown down by the feeble current which is excited.

Fig. 37.

105. When a solution of acetate of lead is decomposed, a very beautiful result ensues at the positive pole. To exhibit this, a large polished metallic surface is connected with the positive wire in the solution, and a fine negative pole is brought near to it. Metallic lead is precipitated on the negative wire, and at the same time a precipitate of deutoxide of lead takes place on the positive plate, exhibiting concentric rings of prismatic colors, owing to the different thickness of the coating of oxide. These depositions have received the name of *metallic chromes.* In this instance, the oxygen about to be evolved at the positive pole unites with a portion of the protoxide of lead in solution, and the resulting deutoxide appears on the surface of the plate. Instead of a

fine negative wire, a spherical body connected with it may be approached to the positive plate, the different distances of its parts producing a beautiful gradation of oxide.

III. THERMO-ELECTRICITY.

106. The term *thermo-electricity* expresses a form of electricity developed by the agency of heat. It was discovered by Professor Seebeck, of Berlin, in 1822, that if the junction of two dissimilar metals was heated, an electrical current would flow from one to the other. Thus, if the ends of two wires, or strips of German silver and brass, are made to touch each other, or are brazed together, and the junction heated, a current will flow from the German silver to the brass, if the free extremities of the wires are connected by any conductor of electricity, and an electrical circuit will be established, as the galvanic circuit is established by connecting the poles of the battery. In the cut, Fig. 38, G represents

Fig. 38.

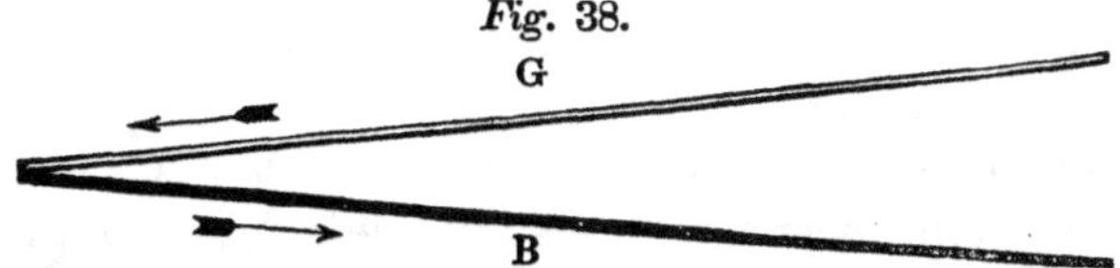

the German silver, and B the brass; the direction of the current being indicated by the arrows.

107. In thermo-electricity, as in galvanism, instead of two metals, one metal, in different con-

ditions, can be used to excite a current. Thus merely twisting the middle of an iron or platinum wire, and heating it on one side of the twisted portion, will produce a current, flowing, at the heated part, from the untwisted to the twisted portion, when the extremities are connected.

108. A current may also be excited with two wires of the same metal, by heating the end of one, and bringing it in contact with the other. It is difficult to succeed in this experiment when metals are used whose conducting power for heat is great. Thus copper or silver wires produce a very feeble current; but iron or platinum an energetic one, especially when the ends, which are brought in contact, are twisted into a spiral. The direction of the current at the junction is from the cold to the hot wire; and it ceases as soon as an equilibrium of temperature is established between the two. A considerable current is also produced by heating the junction of two platinum wires of different diameters. The current flows from the fine to the coarse wire, whether the heat is applied at the point of junction or to either wire at a little distance from it. In large arrangements, plates or strips of dissimilar metals are generally used.

109. The thermo-electric current, thus excited between two metals, is generally referred to the difference in their conducting power for heat, and to the different orders of crystallization to which their particles belong, the laws of crystallization being supposed to result from the electrical character of the

particles. Where the same metal, in different conditions, is used, the production of electricity is referred to the unequal propagation of heat on each side of the heated point, caused in the single wire by the obstruction occasioned by the twist, and in the case of two wires, by the contact of the cold wire, or where they are connected together, by the difference in their diameters. The causes, however, have not yet been fully investigated, and many points are involved in great obscurity.

110. Metals differ greatly in their power to excite a current, when associated together in thermo-electric pairs. Some of the peculiarities in the combinations of the more useful metals will be stated. It is necessary, however, to say a few words with regard to the *galvanometer*, an instrument to indicate or measure electrical currents, and which will be more fully described in Book I. Chapter 2. A current of electricity, passing through a wire, is found to deflect a magnetic needle in its neighborhood. By an

Fig. 39.

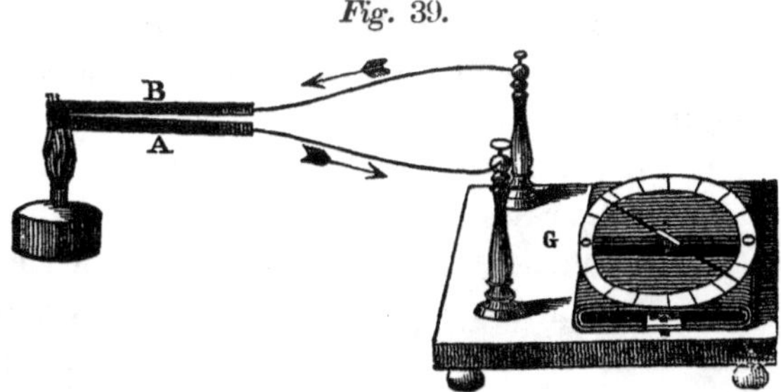

arrangement such as Fig. 39, where G is the galvanometer, consisting of a magnetic needle in close proximity to a coil of wire, above which is fixed a graduated circle, the direction of an electrical current made to pass through the wire is indicated by the deflection of the needle from the north and south line, in one direction or the other, and its strength is measured by the number of degrees to which it is deflected. The *deflection of the needle* will be frequently referred to hereafter. When using this instrument, the zero points of the graduated circle must be placed north and south, so as to lie in the same direction as the magnetic needle above them does when influenced by the earth's magnetism alone. In the figure, a thermo-electric pair, of bismuth and antimony, heated by a spirit-lamp, is shown in connection with the galvanometer. The arrows indicate the course of the current from the antimony, A, to the bismuth, B, in the exterior circuit; its direction being of course the reverse of that at the junction, where it flows from B to A.

111. The character of the juncture between the plates or wires has an important influence on the amount of the current with the same metals. Frequently, when the elements of the pair are merely made to touch each other, the current is greater than when they are brazed or soldered together. Generally, the slighter the connections are, the better. They must be sufficient to conduct all the electricity generated, but no more; for if they are unnecessarily large, they allow the electricity to return to the metal whence it proceeded, without passing through the exterior circuit.

112. The metal from which the current proceeds through the heated junction is analogous in situation to the zinc, or positive plate, in the galvanic pair, from which the current proceeds through the liquid of the battery, as described in § 16. The metal to which the current proceeds, through the junction, is analogous to the copper, or negative plate. The positive pole of the thermo-electric pair is the extremity of the negative metal, as the copper pole is the positive pole of the battery. The negative thermo-electric pole is the extremity of the positive metal. In the observations and table which follow, the positive element of the pair, answering to the zinc in a galvanic pair, will always be named first.

113. German Silver and Antimony. — The current excited by these is greater than that from bismuth and antimony at the same temperature. Their junctions being put into hot oil, of a fixed temperature, and the free ends of the plates connected with the galvanometer used in these experiments, the bismuth and antimony occasioned a constant deflection of the needle of 75°; the German silver and antimony, a deflection of 85°: the heat being increased with the bismuth and antimony to the melting point of bismuth, the deflection was 82°, while the German silver and antimony, heated in a spirit-lamp, gave a deflection of 88°.

114. Bismuth and Antimony. — Plates of these metals have been heretofore generally used in large thermo-electric arrangements. The current excited by heating their junctions is greater than from many

other metals, when a feeble heat is used; but from the fusibility of bismuth, the heat can never be raised very high.

115. German Silver and Carbon.—A current of considerable energy is produced by this combination. In this and in the succeeding experiments, where the use of carbon is mentioned, the kind employed was the gas-carbon deposited from carbureted hydrogen in the retorts of the gas-works. This substance lines the upper part of the interior of the retorts in compact layers, and is entirely different from coke, which is the residue from the coal, after its gas has been expelled. It is nearly or quite pure carbon, and is a better conductor, both of heat and electricity, than ordinary charcoal.

116. German silver is an alloy of nickel with copper and zinc, containing in 100 parts, 50 of copper, 30 of zinc, and 20 of nickel. It is used in many of the thermo-electric instruments, which will be hereafter described. German silver is positive to all the metals that have been tried, even to nickel itself; with the exception of bismuth, to which it is negative.

Carbon and Silver, or Iron.—In these combinations, and also with antimony, the carbon is positive, the current being rather feeble.

117. The deflections given in the following table admit of comparison with each other to a considerable extent, though not so strictly as if wires of the same size had been employed in all the experiments. It must be remembered, too, that as the needle

approaches the extreme angle of deflection, 90°, a much greater increase of the current is required to carry it a few degrees farther towards 90° than when it is near the zero. Hence, a deflection of 40° does not indicate a current of half the power of one of 80°, but considerably less. Nor can momentary deflections be compared with permanent ones, in estimating the power of the current; as a current which, by its first impulse, causes the needle to traverse a large arc, may not be able to maintain more than a few degrees of steady deflection.

Current flows through Heated Junction. *From positive*	*To negative.*	Deflection of the Needle.
German Silver, ...	Antimony, ...	88°
German Silver, ...	Silver,.	85°
German Silver, ...	Brass, ...	85°
German Silver, ...	Iron,.	85°
German Silver, ...	Palladium, ...	85°
German Silver, ...	Copper, ...	85°
German Silver, ...	Cadmium, ...	85°
German Silver, ...	Zinc, ...	84°
German Silver, ...	Platinum, ...	81°
German Silver, ...	Carbon, ...	82°
Silver, ...	Antimony, ...	88°
Bismuth, ...	Antimony, ...	82°
Bismuth, ...	Silver, ...	78°
Bismuth, ...	Palladium, ...	85°
Bismuth, ...	Carbon, ...	85°
Bismuth, ...	German Silver, ...	83°
Platinum, ...	Carbon, ...	78°
Carbon, ...	Antimony, ...	75°

118. The actual, though not the relative, amount of the deflections, will vary with the galvanometer

employed. In accurate experiments, the arrangement should be such that the deflections shall not exceed 20°. Within this limit, the deviation of an astatic needle from the zero point is strictly proportional to the quantity of electricity circulating.

119. The wires were not soldered together, but their ends were brought in contact before the application of the heat, and kept so to the end of the experiment. With the more fusible metals, the greatest heat was employed which was consistent with their fusibility. The object was to produce the greatest current that could easily be obtained from each combination. It will be found that there is an entire difference between the series of positive and negative metals for thermo-electricity and for galvanism.

120. In some cases, the direction of the current is reversed, either by raising the heat at the junction to a high degree, or by heating one metal more than the other. The following are instances of this kind. The metal of each combination, which is positive at low temperatures, is named first. Increasing the temperature of the negative metal generally increases the amount of deflection produced by heating the junction; while, if the higher heat is applied to the metal which is positive at moderate temperatures, a current in the opposite direction is established. The direction of the current in these combinations is, however, often uncertain, and the few experiments which have been made afford no explanation of the cause of the changes.

121. Iron and Platinum. — When heat is applied to the junction, or to the platinum a little one side of it, a deflection of about 50° is obtained; when to the iron near the junction, or when the junction itself is raised to a red heat, the direction of the current is immediately reversed, it now flowing from the platinum to the iron, and the needle is deflected 60° or 70° in the opposite direction.

122. Copper and Iron. — With fine wires the current is feeble, with large ones tolerably powerful. The deflection is increased by heating the iron near the junction. When the junction is raised to a red heat, the current is reversed, and still more readily when the heat is applied to the copper near it.

Silver and Iron. — Deflection considerable. On heating the silver, an energetic current ensues in the opposite direction; also, in a less degree, by raising the junction to a red heat.

Brass and Iron. — Current moderate; reversed at a red heat, and still more effectually by heating the brass.

Zinc and Iron. — Current moderate, and on heating the zinc near the junction to its melting point, changes its direction.

123. Platinum and Silver. — Deflection 70°. On heating the platinum, a strong current flows in the opposite direction.

Brass and Silver. — The current is reversed at a red heat, or by applying the heat to the brass, near the junction.

124. In *quantity*, the thermo-electric current much

resembles a feeble galvanic current. In *intensity*, it is somewhat less. In a single galvanic pair, electricity is set in motion in a certain direction, and cannot return in the same path to the zinc, from which it proceeded, without passing through the fluid between the plates, which is a poor conductor. It is, therefore, partially, though very imperfectly, insulated. In a thermo-electric pair, the electricity is set in motion from one of the metals to the other, through the metallic junction. Here there is no insulation. The current flows through a good conductor, and can only be the excess of the force which sets the electricity in motion over its constant effort to return to equilibrium. It is probably for this reason that the intensity of thermo-electricity is less than that of galvanism.

125. A single galvanic and thermo-electric pair were taken, each of which deflected the needle 75°, permanently. The galvanic current was then made to flow through a hundred feet of fine steel wire 1–150 of an inch in diameter. From the poor conduction of the wire, the deflection was reduced to 60°. With the thermo-electric pair, it was found that the needle was deflected 60°, when the current passed through only fourteen feet of the wire. As the conducting power of a wire is in proportion to the intensity of the current, the intensity of the thermo-electric current in this instance was equal to one seventh of that of the galvanic current.

126. In soldering the wires or plates together, they are not usually connected in a straight line,

but at an acute angle with each other. If several of these single pairs are associated together consecutively, that is, by connecting the German silver of the one to the brass of the next, or the bismuth of one to the antimony of the next, and so on, we have a thermo-electric battery, in which the power of the current is exalted as by the successive pairs of plates in a galvanic battery. It will be understood that in these cases there is German silver and brass alternately, or bismuth and antimony alternately, &c., throughout the whole series. For the sake of compactness, the wires or plates are laid side by side, and soldered by their alternate ends, while they are insulated or separated from each other by paper or pasteboard, which prevents any passage of electricity from one to the other, except through the junctions.

127. A thermo-electric series of considerable power may be constructed of strips of German silver and brass. It will bear contact with red-hot iron, and is very compact. Strips of pasteboard are interposed between the adjacent metals. Fig. 40 represents such a series, consisting of ten pairs. When several

Fig. 40.

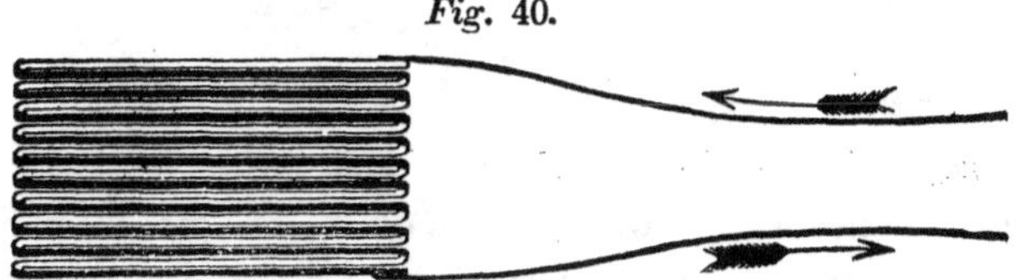

pairs are connected in this manner, it is necessary that the junctions should be somewhat larger than in the case of a single pair. Then, the slighter the connection the better; but as the current has to flow through all the junctions in a series of pairs, the electricity generated would scarcely be conducted through them at all, were they all imperfect. By heating the junctions of the strips at one end of the series with a spirit-lamp, a current is produced which increases or diminishes as the heat is applied, depending altogether for its existence on the difference of temperature in the opposite junctions. On grasping the junctions, at one end, in the fingers, even the warmth of the hand produces a sensible effect. It is evident that, if the junctions at both ends of the series were equally heated, currents would be produced in opposite directions, which would neutralize each other.

128. Thermo-Electric Battery.—In Fig. 41 is represented a thermo-electric battery, composed of

Fig. 41.

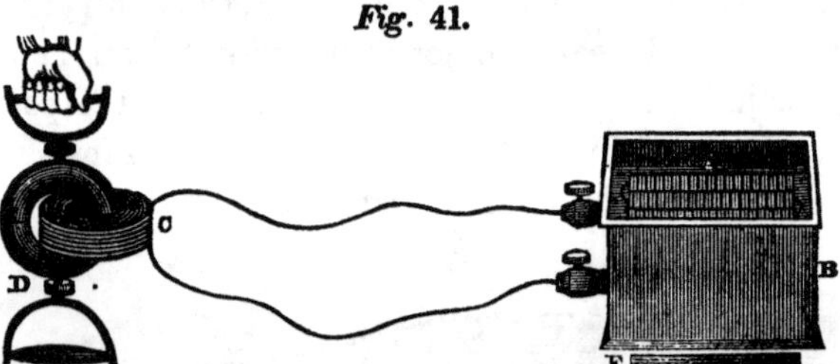

sixty pairs of plates of bismuth and antimony, each three inches long, three fourths of an inch broad, and one fourth of an inch thick. They are arranged side by side, in a metallic case, B, so that one series of junctions, underneath the battery, may be heated by the radiation of a hot iron plate, E, the edge of which is seen underneath the battery in the cut; while the opposite junctions, seen at A, are kept cool by water or ice placed in the receiver, which forms the upper part of the battery. A still greater depression of temperature is produced by a mixture of snow or pounded ice, with half its weight of common salt. In order to make the receiver sufficiently water-tight, as well as to insulate the plates from the case, they are cemented into it with plaster. Refrigeration at one end of the bars, as would be anticipated, is found to produce a current in the same direction, and equal to that which would be produced by a similar excess of heat at the other end; difference of heat at the different ends, however produced, being the occasion of the current. By associating both of these causes in this battery, there is a corresponding increase of power. As the metals employed in the battery are fusible, the radiant heat of the iron ought never to exceed 300° Fahrenheit. The iron plate being laid upon a large tile, the battery is placed over it, the iron being pretty near the ends of the bars, but not in contact with them.

129. The terminal plates of the battery are connected with two screw-cups, passing through the metallic case, and insulated from it. In the cut, the battery is seen in connection with an apparatus, which will hereafter be described, by which the magnetizing power of the current is shown. The ends of the coil of insulated wire, C, being fixed in the cups, the current is obliged to traverse the coil, and the two semicircular bars of soft iron, seen at

D, are held together, by the magnetism thus induced, with so much force as to require a weight of forty or fifty pounds to separate them. This battery has sufficient power to give shocks and sparks, and produce various magnetic phenomena, with the appropriate apparatus, which will be described hereafter, when the principles on which those effects depend have been explained.

130. By forming a compound thermo-electric battery of a number of slender bars of antimony and bismuth, soldered together by their alternate ends, an instrument is obtained far more susceptible to slight differences of temperature than the mercurial or air thermometer. Such is the construction of Melloni's *Thermo-Multiplier.* It consists of about fifty little bars of the two metals, enclosed in a brass cylinder, one half or one third of an inch in diameter, and two and a half inches long. The terminal bars or poles of this little battery are connected by wires with a delicate galvanometer. When the ends of the bars at one extremity of the cylinder are exposed to any source of heat, while the temperature of the other ends remains unchanged, an electric current is excited which deflects the needle of the galvanometer more or less in proportion to the heat. The radiant heat from the hand brought near the battery is sufficient to move the needle several degrees. This instrument was employed by Melloni in his important researches on the transmission of radiant heat from different sources through various transparent and translucent bodies.

131 In thermo-electricity, an electrical current is produced by heating unequally the opposite ends of metallic plates, associated in a thermo-electric series. The converse of this is found true. If a galvanic current is made to pass through the same series, the opposite junctions will acquire heat on the one side and lose it on the other.

132. Fig. 42 represents an instrument for showing the simultaneous production of heat and cold by the galvanic current. It consists of three bars, two of bismuth and one of antimony, arranged as seen in the figure, where the antimony is shown at A, and the two bars of bismuth at B and B′. The bars are soldered together under the bulbs of two air thermometers, T and T′, a little cavity being made to receive the bulb of each thermometer; a drop of water is put in each cavity, in order to facilitate the conduction of heat from the metals to the thermometers. The galvanic current being sent through the metals, in the direction indicated by the arrows, from the bismuth B′, through the antimony, to the other bar of bismuth, and thence back to the battery, cold is produced at the junction of A with B′, as will be indicated by the thermometer T′, and heat at the junction between A and B, as the thermometer T will show. By reversing the direction of the battery

Fig. 42.

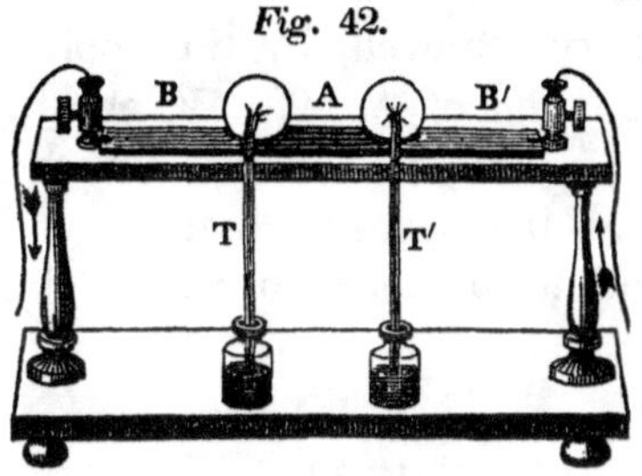

current, the effect on the two thermometers will be reversed.

133. The elevation of temperature produced is always greater than the depression; this difference is probably due to the low conducting power of the metals for electricity, which causes them to become slightly heated by the current — a phenomenon altogether distinct from the heating of the junction by it. It will be observed in the figure that the current has the same direction as that which would be produced, were the battery removed, by the application of heat at the junction of A with B′, or of cold to that between B and A; the current which produces heat flowing in the opposite direction to the current which would be produced by it.

IV. ANIMAL ELECTRICITY.

134. This name has been applied to the electricity developed in the organs of certain fishes. These organs are found, by anatomical examination, to consist of numerous cells, arranged in columns, and divided by septa or small discs, so as to suggest at once the idea of a galvanic arrangement. Further observations show that they are supplied with large branches from the nervous system, so much so, that, in the torpedo and gymnotus, the ganglia and nerves belonging to the electrical apparatus exceed in extent all the rest of the nervous system of the animal. It is also found that this manifestation of electrical

power is subject to the nervous system, and controlled by the will. An actual connection, therefore, exists here between the vital and electrical phenomena. Several cases have recently been authenticated, in which electricity has been evolved by the human subject in a state of nervous disorder. The animals naturally endowed with this power are surrounded by a conducting medium, water, which enables them to paralyze other creatures in their neighborhood, which thus become their prey.

135. There is a remarkable analogy between this production of electrical force and the production of muscular force in the ordinary contraction of the muscles. In both cases, the power which is developed is inherent in the organ where it is manifested, and in both the power is controlled and exercised by the agency of the will or influence of the nervous system. Liebig has suggested a theory of muscular action, which supposes that the contractile force of the muscles is due to a principle set free by the oxidizement of the animal tissues by the blood, in the same way that electricity is set free by the oxidizement of zinc. He supposes that, when a muscular contraction takes place, the nerves, supplying the part, withdraw their vital protection, and oxidation under the chemical laws, and the consequent development of force, result. An evidence relied upon in support of this theory is the waste of animal substance which occurs proportional to muscular action. A further application of this theory would extend it to the electrical organs of fishes, where,

by oxidizement of tissues, suitably arranged, electricity itself, instead of muscular force, would be eliminated, when the protecting agency of the nervous system was withdrawn. An illustration of the supposed active and passive states of the animal organs under the nervous control may be readily found in Smee's battery, where the plates remain opposed to each other without any action until the circuit is completed, when an electrical discharge at once takes place. However hypothetical the views of Liebig may be, they may be useful as connecting together some phenomena belonging to a very dark and unexplored field of science. The analogy of the action of the will or nervous system in producing, in the one case, muscular, and in the other electrical action, is at any rate an important one. In the diseased human subjects which have been mentioned, the development of muscular force seems to have been changed into a development of electricity. We have, as yet, no knowledge of the relation between these two principles. Professor Renwick has, however, remarked, "In these electric fish we behold nervous power converted into electric force; it cannot be doubted that the converse of this is possible."

136. The Gymnotus is found in the fresh waters of South America. It is interesting as having been the subject of a series of experiments by Faraday, in which the phenomena of magnetism, electrical decomposition, the production of heat, the shock and spark were all obtained. Fig. 43 represents an individual of this species, which was preserved alive for

some time in this city. The upper figure gives a lateral view, the lower one a view from above, with the hand in a favorable position to receive a shock, passing from the anterior to the posterior part of

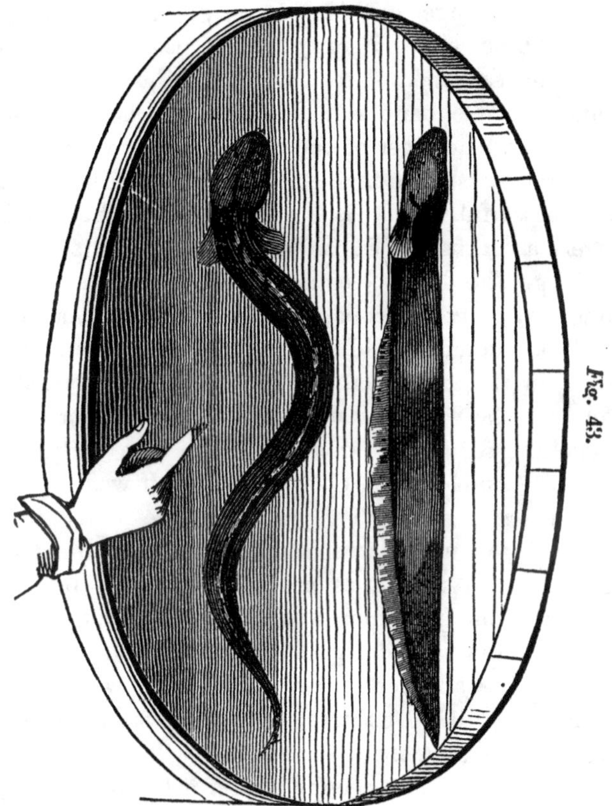

Fig. 43.

the animal, through the surrounding water. It will readily be comprehended, that, even when the animal is in a straight position, the finger placed near its side will become the medium of the passage of electricity, as it is a better conductor than water, and the current between the extremities extends to some distance from the animal. The electricity passes through the finger in this case, only from one side to the other, but still, by acting on the local nerves, produces the sensation of a considerable shock. When a finger of one hand is placed near the head of the animal, and a finger of the other near the tail, the discharge passes through the arms and body, and a very powerful shock is experienced. When two fingers are thus placed in water, in the course of a discharge of ordinary electricity, a shock is felt in the same way; but in proportion as the water is made a good conductor by adding salt, the effect decreases. Thus, in the experiment (§ 52) with a leech or fishworm, if the water be made a good conductor, the animal ceases to be troubled when the galvanic circle is completed. When a small fish was placed in the water with the gymnotus, the latter was observed to curve himself so as to bring the fish between the extremities of his body, and then to make a discharge, which was either fatal or produced entire numbness and rigidity.

137. The anterior region of the body was found by Faraday to be positive, and the posterior negative, at the moment of discharge. The electrical organs of the animal occupy almost its whole length, with the exception of the anterior region. It is provided with a beautiful fringe-like ventral fin, which gives it great ease of motion. It is found to suffer

great exhaustion from the frequent exercise of its electrical power, showing thus another link between the nervous and electrical systems.

138. The, TORPEDO, or ELECTRICAL RAY, is found on the sea-coast in different parts of Europe. A new

Fig. 44.

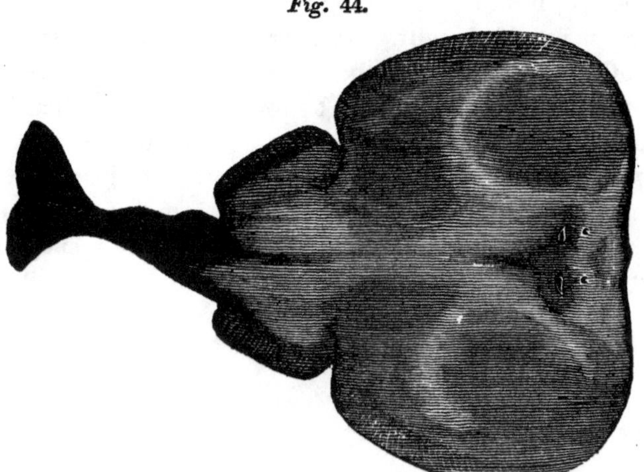

species *Torpedo occidentalis*, Storer, has lately been taken at Wellfleet, Massachusetts. It is represented in the above figure. They are described as varying in weight from 20 to 200 lbs., and the shocks which they give, according to the testimony of fishermen, are sufficient instantly to prostrate a man These shocks are also felt, in a less degree, by one holding a harpoon or rope, to which the animal is attached. The electrical organs are situated on the anterior and lateral portion of the animal. The electrical lobes of the brain are larger than the whole remainder of that organ, and the volume of the electrical nerves greater than those supplying all other portions of the animal.

139. The Silurus Electricus, the third of the electrical fishes, is found in the rivers of Africa. In shape, it is elongated like the majority of fishes, and is provided with cirri, or feelers, about the mouth. It is believed that the fishes of this class remain concealed near the bottom, and decoy their prey by waving these cirri, so as to counterfeit the appearance of worms. This would give to the electrical Silurus a great advantage in bringing other animals within the scope of the powerful agency with which it is endowed.

MAGNETISM.

I.

DIRECTIVE TENDENCY OF THE MAGNET.

I. IN REFERENCE TO ANOTHER MAGNET.

140. Attractions and Repulsions. — The effects produced by the opposite poles of a magnet, though in some respects similar, are in others contrary to each other; the one attracting what the other repels. Poles of different magnets, of the same name, that is, both north or both south, are found to repel, while those of an opposite name attract each other. This is shown by the following experiment.

Let the south pole of a magnet, N S, be brought near to the north pole of a dipping needle, mounted upon a stand, as represented in Fig. 45. The needle, previously nearly vertical, will move into the position *a a*, its north pole (indicated by the head of the arrow in the cut) being attracted by

Fig. 45.

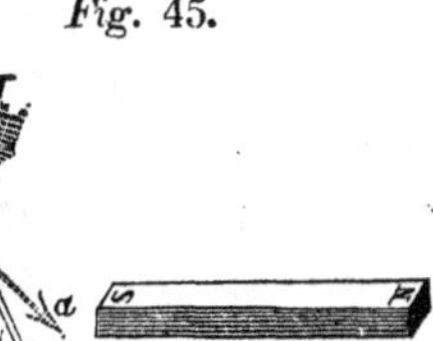

the south pole of the magnet. If, now, the magnet be reversed, and its north pole be made to approach the corresponding pole of the needle, the latter will assume the position *r r*, its north pole being repelled by that of the magnet. The south pole of the needle, on the contrary, will be *repelled* by the south pole of the magnet, and *attracted* by its north pole.

141. In place of the dipping needle, a compass needle, or any horizontal magnetic needle, can be employed. A very fine and perfectly dry sewing-needle, being previously magnetized, and then laid carefully upon the surface of water, will float, and, being thus at liberty to move freely in any direction, may be conveniently used to show the attractions and repulsions described above. A larger needle will answer equally well, if passed through a small piece of cork, to enable it to float.

142. The intensity of the attraction or repulsion exerted between two magnetic poles varies in the inverse ratio of the square of their distance; that is, if the distance of the poles is doubled, the force with which they attract or repel each other is reduced to one quarter of its previous amount; if their distance is trebled, to one ninth; and so on. These attractions and repulsions are not affected by the interposition of glass or metal, or any substance whatever between the two magnets; unless the interposed body is itself susceptible of magnetism.

143. In consequence of the mutual attractions and repulsions exerted between the poles, a magnetic needle, brought near to a fixed magnet, assumes cer-

tain determinate directions depending upon the relative positions of the two. Thus, in Fig. 46, N S represents a stationary bar magnet, and the three arrows indicate the directions taken by a magnetic needle placed in different positions. The two needles near the south pole of the bar direct their north poles towards it, the north pole of the stationary magnet being too remote to influence them to any extent. But the needle placed over the centre of the bar has its direction determined by both poles, and assumes a position parallel to the bar, but with its poles in a reverse direction.

Fig. 46.

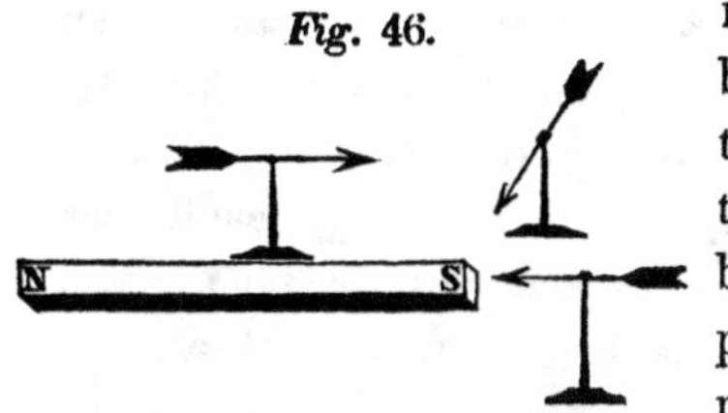

144. In Fig. 47 are represented various directions assumed by a small movable magnet brought near to a stationary one. These may be readily observed by carrying a small bar magnet, suspended by a string, or, better, a magnetic needle, mounted upon its stand, slowly round a stationary bar magnet.

Fig. 47.

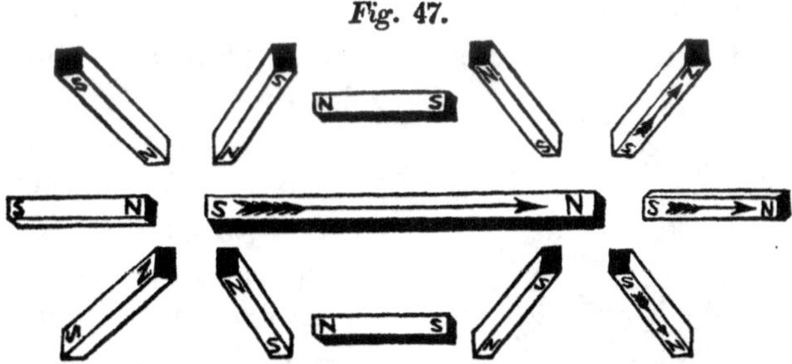

145. The arrangement of iron filings around the poles of a U-magnet, as represented in Fig. 48, depends upon the directive tendency communicated to the minute particles. As will be explained hereafter, each particle of the filings becomes a temporary magnet, with a north and a south pole, while attached to the steel magnet. The filings are held together by their mutual attractions, and, having thus great freedom of motion, readily obey the directive influence exerted upon them.

Fig. 48.

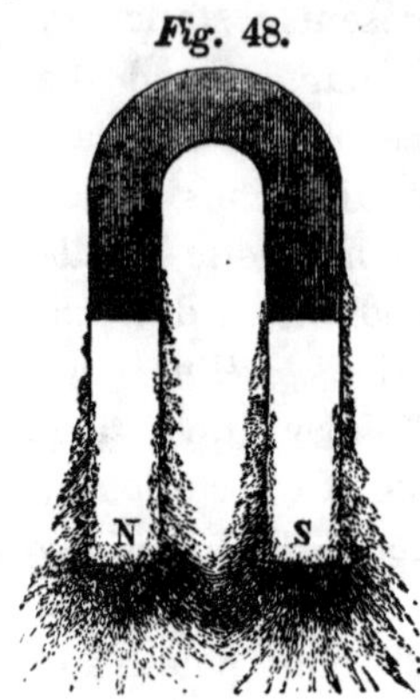

146. Spread a thin covering of iron filings over a sheet of paper, or thin pasteboard, and place a powerful U-magnet vertically beneath it, with the poles close to the paper. The dotted lines in the cut (Fig. 49) show the arrangement which the particles of iron will assume. Each one becomes a magnet with two poles, and connects itself with those adjoining it, so as to form curved lines of a peculiar character. This experiment may be performed n a still more satisfactory manner, by supporting the paper, with the magnet in contact with its under surface, and then showering down iron sand or iron

Fig. 49

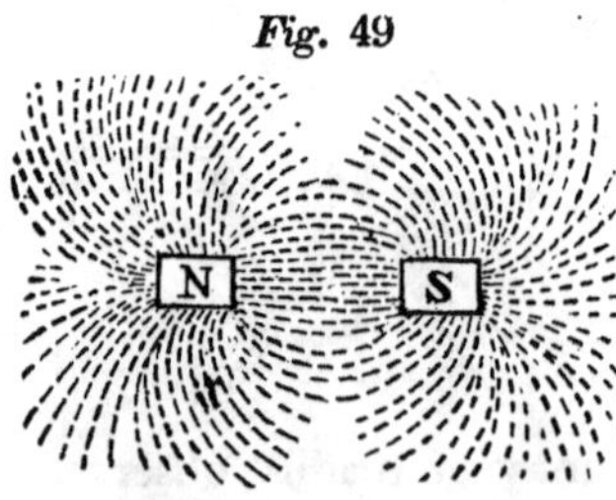

filings from a sand-box held some inches above. The particles of iron, as they strike the paper, can thus more readily assume the positions to which they tend under the magnetic influence.

147. The lines formed by the filings afford a good experimental illustration of what are called *magnetic curves;* that is, the curves into which an infinite number of very minute magnetic needles suspended freely would arrange themselves, if placed in all possible positions about a magnet. When the particles are very small, the *attractive force* exerted upon them by the magnet, being the difference of its action upon the two poles of each particle, is slight; while the *directive force* is very considerable. The direction assumed by each particle, and consequently the form of the magnetic curve, connecting any point on one half of the magnet with the corresponding point of the other half, is deducible, on mathematical principles, from the laws of magnetic attraction and repulsion. The curvature of the lines is due to the combined action of the two poles of the magnet. If only one pole acted on the minute particles, they would arrange themselves in straight lines, diverging in all directions from the pole, like radii from the centre of a sphere. This may be partially shown, by placing a bar magnet perpendicularly under the paper which is strewed with filings, with its upper pole close to the sheet.

II. IN REFERENCE TO A CURRENT OF ELECTRICITY

148. It was discovered by Professor Œrsted, of Copenhagen, in the year 1819, that a magnet, freely suspended, tends to assume a position at right angles to the direction of a current of electricity passing near it. This may be shown by the following experiment: —

149. Let N S (Fig. 50) be a magnetic needle poised upon a pivot so as to allow of a free horizontal motion, and W R a wire passing directly over and parallel to it. Of course, the direction of the wire must be north and south, as the needle will necessarily assume that direction, by the influence of the earth. If, now, the extremities of the wire are put in connection with the poles of a galvanic battery, in such a manner as to cause a current of electricity to pass through it, the needle, N S, will be deflected, and will turn towards the position *a b* or *c d*, according to the direction of the current of positive electricity, whether from W to R, or from R to W. If the wire is placed in the same direction below the needle, the deflections will be the reverse of those produced by

Fig. 50.

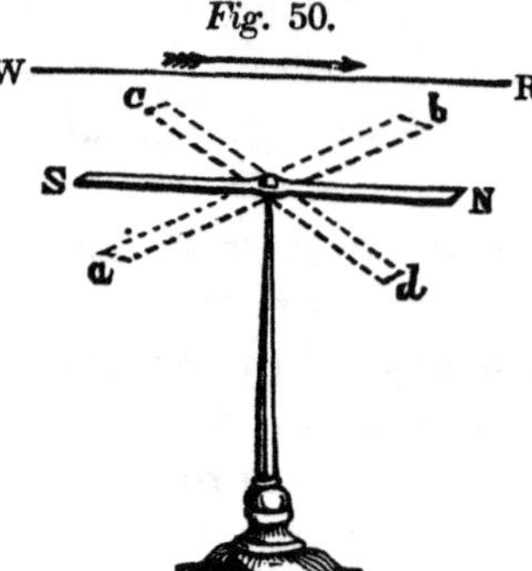

the same current when flowing above. If the positive current is passing from south to north in the wire, as shown by the arrow in the cut, the north pole of the needle will turn to the west if it be below the wire, and to the east if above it.

150. In these cases, the needle will not be deflected so far as to assume a position exactly at right angles with the wire, on account of the influence of the earth, which still acts upon the magnet, and tends to draw it back to its original position. It will accordingly come to rest in a state of equilibrium between the two forces, in a direction intermediate between a line at right angles to the wire and that of the needle when controlled by the magnetism of the earth alone.

151. The same experiment may be performed with the dipping needle, the wire being placed parallel with the needle. By thus varying the positions of the wire and the needle, it will be found that in all cases the needle tends to place itself at right angles with the wire, and to turn its north pole in a determinate direction with regard to the wire.

152. The action of a conducting wire upon a magnet exhibits, in one respect, a remarkable peculiarity. All other known forces exerted between two points act in the direction of a line joining these points; such is the case with the electric and magnetic actions separately considered. But the *electric current* exerts its magnetic influence laterally, at right angles to its own course. Nor does the magnetic pole move either directly towards or directly from

the conducting wire, but tends to revolve around it without changing its distance. Hence the force must be considered as acting in the direction of a *tangent* to the circle in which the magnetic pole would move. It is true that, in many positions of the magnet with regard to the wire, apparent attractions and repulsions occur ; but they are all referable to a force acting tangentially upon the magnetic poles, and in a plane perpendicular to the direction of the current. This peculiar action may be better understood by means of a figure.

153. Thus, let *p n* (Fig. 51) be a wire, placed in a vertical position, and conveying a current downwards (*p* being connected with the positive pole of the battery.) Now, suppose the north pole of a magnet, N, to be brought near the wire, and to be perpendicular to any point C. If free to move, the pole will revolve around C as a centre, in the direction indicated by the arrows in the cut ; that is, in the same direction as that of the hands of a watch, when its face is upwards. The plane of the circle which the pole describes is horizontal. On causing the current to *ascend* in the wire, the pole will rotate in the opposite direction. If the wire is placed in a horizontal position, the plane in which the pole revolves will, of course, be vertical. The actions of either a de-

Fig. 51.

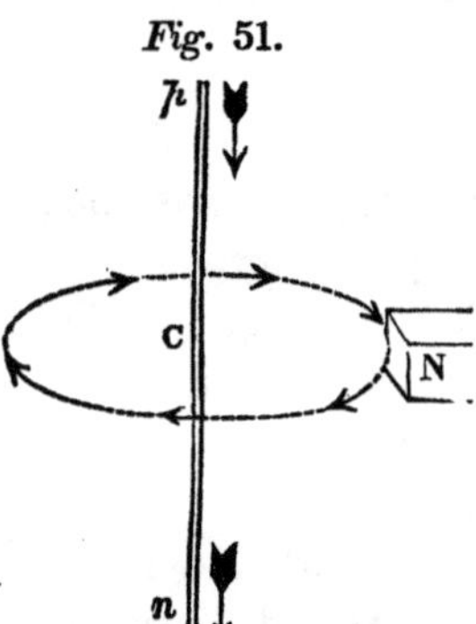

scending or an ascending current upon the south pole are exactly the reverse of those exerted on the north pole.

154. In the experiment given in § 149, no revolution occurs, because the current, acting at once on both poles, tends to give them motion in opposite directions; so that the magnet comes to rest in a position of equilibrium between these two forces, across the wire. It will be shown hereafter that a continued rotation may be produced by confining the action to one pole. If the wire is movable and the magnet fixed, the former will revolve around the latter in a similar manner, and in the same directions. Thus a wire conveying a descending current tends to rotate round the north pole of a magnet, in the direction of the hands of a watch.

155. The two following articles of apparatus are used to show the directive tendency of a magnet in reference to an electric current.

156. ASTATIC NEEDLE. — A magnetic needle, so contrived that its directive tendency in respect to the earth is neutralized, so that it shall remain at rest in any position, is called an *astatic needle.* It is constructed as represented in Fig. 52, consisting essentially of two needles, one above the other, placed in positions the reverse of each other in respect to their poles Such a system will not be affected by the

Fig. 52.

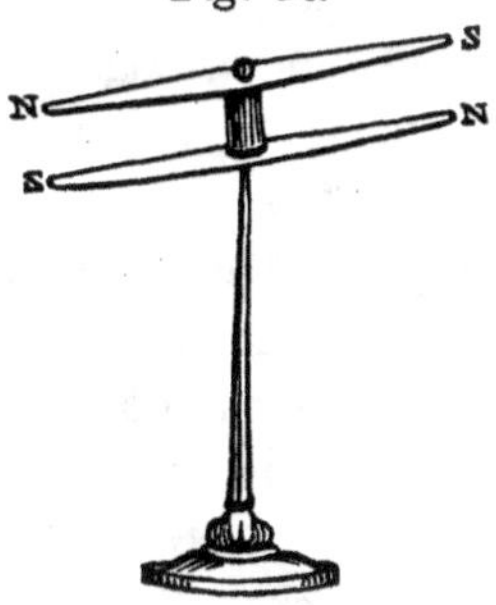

S N N S

magnetic influence of the earth, as whatever forces may be exerted upon the upper needle will be counteracted by equal forces exerted in reverse directions upon the lower. It would be the same, indeed, with the influence exerted by the current of electricity, if the wire could be placed in such a position as to act equally on both needles. But by placing the wire parallel to and above the upper needle, the influence of the wire will be far more powerful upon the upper than upon the lower one, and, the action of terrestrial magnetism being neutralized, the needle will assume a position exactly at right angles with the conducting wire. If the wire be placed as nearly as possible between the needles, and parallel to them, the influence of the upper side of the wire will deflect the upper needle in the same direction as the lower needle will be deflected by the action of the lower side of the wire, causing a more powerful effect.

157. Magnetic Needle, half brass.—In this instrument the steel needle is wholly upon one side of the point of support, and is counterpoised by a brass weight on the other side. By this arrangement, the action of a current upon the pole which is situated at the centre of motion can have no influence in turning the magnet in any particular direction; and its motion will be determined solely by the action upon the other pole; no rotation, however, can be obtained. The object of the instrument is to show the *directive tendency of a single pole* with reference to the electrical current.

158. GALVANOMETERS. — If the wire transmitting the electrical current, after passing over the needle, is bent and returned under it, as in Fig. 53, it might be supposed that, as the electricity which flows from C to A in the upper part of the wire must pass in a contrary direction, in returning from A to B, below (the cup C being connected with the positive pole of the battery, and B with the negative), the influence of the one part of the wire would neutralize that of the other; for it has already been stated that the needle is deflected to one side or the other, according to the direction of the electrical current. And this would in fact be the case, if the returning part of the wire were upon the same side of the needle with the other part, and at an equal distance from it. But a wire transmitting an electrical current, when passing *below* the needle, will produce an effect the reverse of that produced by one passing *above*, if the current in both cases flows in the same direction. Hence, if the direction of the electric current is reversed in the wire which passes below, it will exert a force auxiliary, and not antagonist, to that of the wire passing

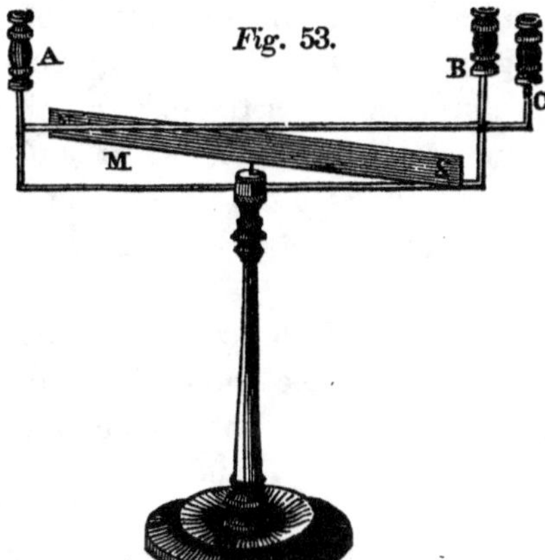

Fig. 53.

above. This is the case with the arrangement here represented. The two portions of the wire are not allowed to touch each other where they cross, but are insulated at that point by some non-conductor of electricity, as by being wound with thread. Instruments of a variety of forms are constructed on the above principle, and are called *galvanoscopes* or *galvanometers*, as they serve to indicate the presence of a current of electricity, and in some degree to measure its quantity.

159. The vertical portions of the wire also aid in deflecting the needle; as may be shown by connecting both the cups B and C with one pole of the battery by two wires of equal length and thickness, and the cup A with the other pole (say the positive). The current will then be divided into two portions very nearly equal, both flowing in the same direction, and at the same distance from the magnet, M, but one below and the other above it. Now, if the horizontal portions of the wire alone acted on the needle, it would remain unaffected; but it will be found to be deflected to a considerable extent by the current which is descending in the vertical portion of the wire near A, and ascending in that below B, as these conspire in their influence.

160. Horizontal Galvanometer. — If the wire is carried many times around the needle, as in Fig. 54, the power of the instrument is much increased, as each turn of the wire adds its influence; provided the wire is not so long, or of so small a size, as to be unable to convey the whole of the current. The

instrument thus becomes a delicate test of the presence of a current of electricity. The coil of wire is

Fig. 54.

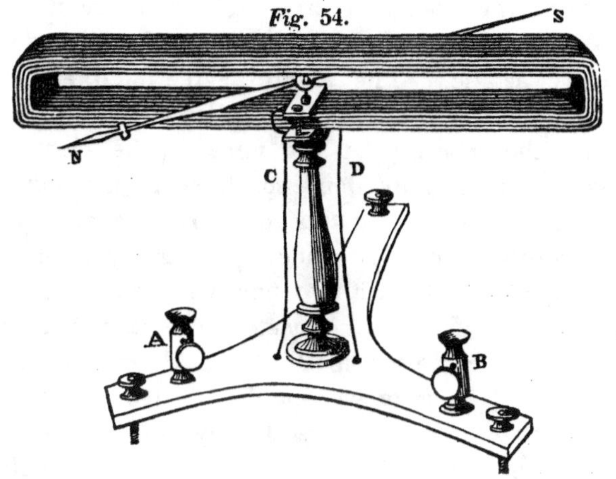

supported on a tripod stand, with levelling screws; the ends, C and D, of the wires being connected with the screw cups, A and B.

Fig. 55.

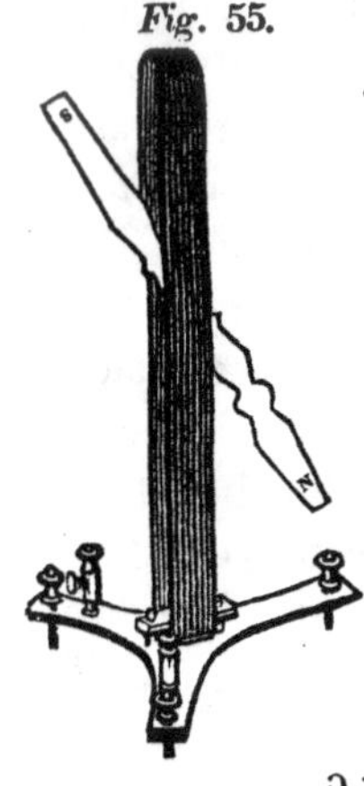

161. Upright Galvanometer.—In this instrument, represented in Fig. 55, both the coil of wire and the needle are placed in a vertical position, the north pole being made a little heavier, in order to keep the magnet perpendicular. When a current is passed through the coil, the deflection is towards a horizontal position. The needle is made of large size, for the purpose of exhibiting the deflections before an audience.

162. Galvanometer with Astatic Needle. — This instrument is represented in Fig. 39, in connection with a thermo-electric pair. It is constructed on a small scale, in order to be delicate; and the needle is nearly astatic. The slight degree of directive tendency which the needle is allowed to retain becomes the measure of the force of the electric current, as the angle of deflection from the north and south line shows how far this resistance is overcome. This instrument may be made so extremely delicate in its indications, that, if two fine wires, one of copper and one of zinc, are connected with it, and their ends slightly immersed in diluted acid, or even in water, it will be very perceptibly affected. Before proceeding to experiment with any galvanometer, it should be so placed that the direction of the coil may coincide with that of the needle, as this is the position of greatest sensibility.

163. The galvanometer is a measurer of what is called the *quantity* of electricity, but takes no cognizance of *intensity*. Mechanical electricity, which possesses great intensity and but little quantity, very slightly deflects the needle of the galvanometer. The current from one galvanic pair influences the needle powerfully, the quantity being very great, and the intensity small. If a hundred pairs be connected together in a single series, the intensity is increased a hundred fold; but the quantity remains the same, and the needle is but little more deflected than by one pair. The reason that there is any difference in this respect is, that, when the electricity is of

high tension, the wire of the galvanometer obstructs the current less, and more actually passes through it.

164. In thermo-electricity, with a single pair, the intensity is less in proportion to the quantity than with a single galvanic pair, and the current is strongly indicated by the galvanometer. The amount of decomposing power in a current of electricity is always exactly as its quantity. The galvanometer indicates, therefore, the electro-magnetic and the decomposing capacity of a current of electricity. An intense electrical current decomposes more easily than one of little intensity; but the amount of matter decomposed is proportional merely to the quantity of the current. Besides the galvanometers in which a magnetic needle is used, other instruments for measuring the quantity of an electric current will be described farther on.

165. When a wire conveying a current of electricity is brought near to a magnetic pole, the pole tends to revolve around it, as has been explained in § 153. If the current acts equally upon both poles, no rotation occurs, because they tend to move in opposite directions; and the magnet rests across the wire in a position of equilibrium between the two forces. But if the action of the current is limited to one pole (which was first effected by Professor Faraday), a continued revolution is produced. If the magnet has liberty of motion, it will revolve around the wire; if the wire only is free to move, it will rotate around the pole. When both the wire and the magnet are at liberty to move, they will revolve in the

same direction round a common centre of motion. A number of instruments have been contrived for exhibiting these movements.

166. Magnet revolving round a Conducting Wire. — In the instrument represented in Fig. 56, the magnet, N S, has a double bend in the middle, so that this part is horizontal, while the extremities are vertical. To its north pole, N, is attached a piece of brass at a right angle, bearing a pivot, which rests in an agate cup fixed on the stand. A wire loop, attached to the upper pole, S, encircles a vertical wire fixed in the axis of motion, and thus keeps the magnet upright. The galvanic current is conveyed by this vertical wire: it is surmounted by a brass cup, A, and its lower end dips into a small mercury cup on the horizontal portion of the magnet. From this part projects a bent wire, which dips into a circular cistern of mercury, open in the centre, to allow the magnet to pass through, and supported independently of it. A wire, terminated by a brass cup, B, for connection with the battery, proceeds outwardly from the cistern. This arrangement allows the current to flow down by the side of the upper pole of the magnet, until it reaches

Fig. 56.

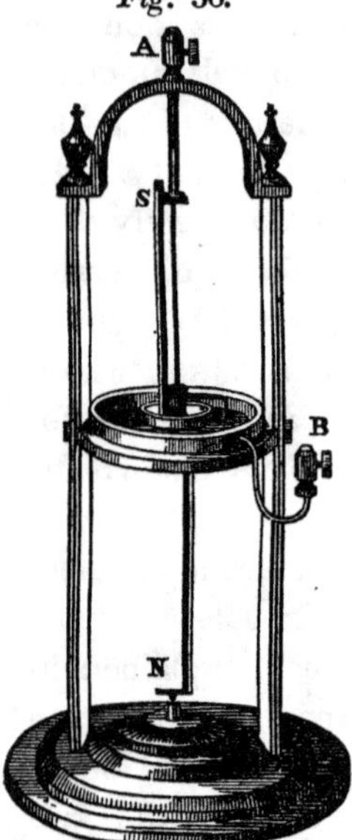

its middle, whence it is conveyed off in such a direction as not to act upon the lower pole. On making connection with the battery, the magnet will revolve rapidly around the wire; the direction of the rotation depending upon that of the current.

167. Magnet revolving on its Axis.—The instrument represented in Fig. 57 is designed to show that the action between the current and the magnet takes place equally well when the magnet itself forms the conductor of the electricity. The lower end, N, of the magnet, being pointed, is supported on an agate at the bottom of a brass cup, connected under the base-board with the screw-cup C. The upper end, S, is hollowed out to receive the end of the wire fixed to the cup A; the brass arm supporting this cup is insulated from the brass pillar by some non-conductor of electricity. To the middle of the magnet is fixed a small ivory cistern, for containing mercury, into which dips the end of a wire, connected with the cup B. Thus the magnet is supported with its north pole downwards, and is free to rotate round its vertical axis. A little mercury should be put into the cavity at S, and into the brass cup at N, and the

Fig. 57.

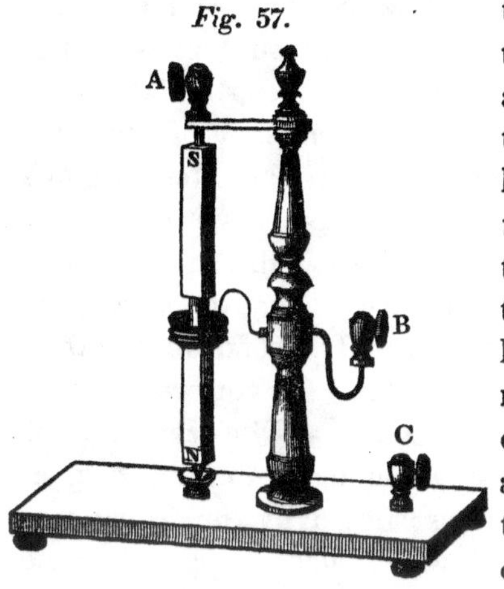

ivory cistern be filled sufficiently to establish a connection between the magnet and the wire attached to B. When the cups A and B are connected with the battery, the current will flow through the upper half of the magnet, causing it to rotate rapidly. If the cups B and C form the connection, the current will traverse the lower half, equally producing revolution of the magnet. Now, connect A and C with the battery, and no motion will result, because the electricity passes through the whole length of the magnet in such a manner that the tendency of one pole to rotate is counteracted by that of the other to move in the opposite direction. Connect B with one pole of the battery, and A and C both with the other pole. The magnet will now revolve; since the current ascends in one half of its length and descends in the other.

Fig. 58.

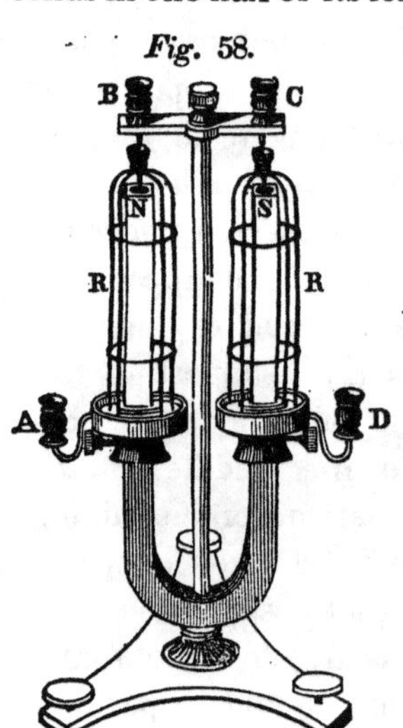

168. Revolving Wire Frame. — The revolution of a conductor round a magnet is shown by the instrument represented in Fig. 58. Two light frames of copper wire, R R, are supported by pivots resting on the poles, N and S, of a steel magnet of the U form; a small cavity being drilled in each pole to receive an agate for the bearing of the pivot. The lower extremities of the wires dip into mercury contained in two circular cisterns sliding

on the arms of the magnet. Bent wires, passing from the interior of the cells, support the cups A and D; and the cisterns themselves are fixed at any required height by means of binding screws attached to them. Each of the wire frames is surmounted by a mercury cup; into these dip the wires projecting downwards from the cups B and C.

169. The cisterns being partly filled with mercury, fix them at such a height that the lower extremities of the wire frames may just touch its surface. The cups surmounting the frames should also contain a little mercury. When the cups A and B are connected with the battery, the left-hand frame will revolve, in consequence of the action of the north pole of the magnet upon the current flowing in the vertical portions of the frame. By uniting C and D with the battery, the other frame will rotate. On transmitting the current from A to D, it will ascend in one frame, and, passing along the brass arm which supports B and C, will descend in the other, causing them both to revolve in the same direction. Instead of the frame, a single wire may be employed, having the form of a loose helix surrounding the pole, its convolutions being a quarter of an inch or more apart.

170. Revolving Cylinder. — This instrument is on the same principle as that last described, and the motion takes place in the same manner; the only difference being that two light copper cylinders, *c c*, Fig. 59, are substituted for the wire frames. These cylinders are serrated at their lower edge, as shown in the figure, to lessen the friction which they expe-

rience in moving through the mercury. The cups for battery connections are lettered in correspondence with those in the preceding cut. (Fig. 58.)

Fig. 59.

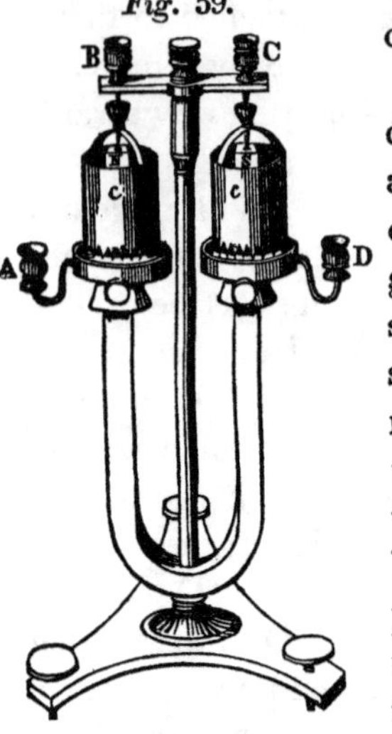

171. In the case of a conducting wire revolving round a magnet, the circumstance of the two being joined together does not affect the result, the wire moving with sufficient power to cause the magnet to turn on its axis with considerable rapidity, when delicately supported: a bar magnet is of course employed. A figure and description of an instrument designed to show this revolution will be found in Silliman's Journal, Vol. XL. p. 111.

172. The current passing within the voltaic battery itself exhibits the same electro-magnetic properties that it does while flowing along a conducting wire connecting the poles. Hence the battery, if made small and light, will revolve by the influence of a magnet. This is effected in the following manner.

173. Rotating Battery.—A small double cylinder of copper, closed at the bottom, is supported upon the pole of a magnet, by means of an arch of copper passing across the inner cylinder, and having a pivot projecting downwards from its under surface, which

rests in an agate cup on the pole. The inner cylinder of course has no bottom. A cylinder of zinc is supported by a pivot in a similar manner upon the copper arch, and, being intermediate in size between the two copper cylinders, hangs freely in the cell. This arrangement allows each plate to revolve independently of the other. In Fig. 60 two batteries are represented, one on each pole of a U-magnet, the one on the south pole being shown in section; in this the zinc plate z is seen suspended within the copper vessel.

Fig. 60.

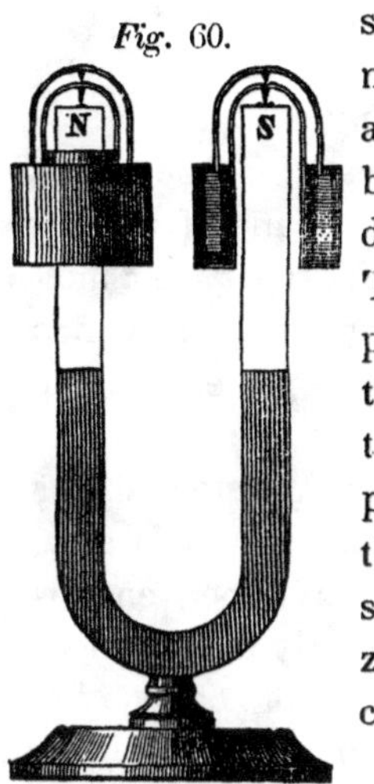

174. On introducing diluted acid into the copper vessel, an electric current immediately begins to circulate, which passes from the zinc to the copper through the acid, and, ascending from the copper through the arch, descends again to the zinc. Hence the zinc plate is in the condition of a conductor conveying a stream of electricity downwards, and will consequently revolve under the influence of the pole which it surrounds. The copper cylinder, on the contrary, is in the situation of a conductor conveying a current upwards, and will rotate in the opposite direction. When there is a battery on each pole of a U-magnet, the two copper vessels will be seen to revolve in contrary directions, and the two zinc cylinders in directions

opposite to these, and of course also contrary to each other.

175. VIBRATING WIRE. — A copper wire, seen at W, in Fig. 61, is suspended over a small basin for containing mercury excavated in the stand, by means of a brass arm supporting a mercury cup, in which the upper end of the wire rests: this mode of suspension allows it to vibrate freely, if its upper end is properly bent. Two cups for connection with the battery communicate, one with the mercury in the excavation, the other with the cup which sustains the wire.

Fig. 61.

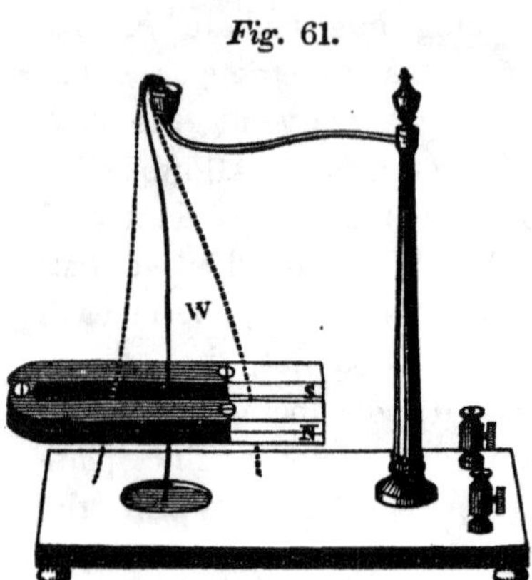

176. The basin being supplied with a sufficient quantity of mercury to cover the lower end of the suspended wire, lay a horseshoe magnet in a horizontal position on the stand, with one of its legs on each side of the wire. When communication is established with the battery, the poles of the magnet will conspire in urging the wire either backwards or forwards between them, according to the direction in which the current flows through it, and the position of the magnetic poles. In either case, the motion will carry it out of the mercury into the position shown by the dotted lines in the cut; and the circuit

being thus broken, the wire will fall back by its own weight; when, the current being reëstablished, it will again quit the mercury, as before, and a rapid vibration will be produced. The vibration may be made somewhat more active by raising the magnet a little from the stand, and nearly to the height of the middle of the wire.

177. The instrument represented in Fig. 62 differs from the one last described in the mode of suspension of the wire and the vertical position of the magnet. The wire is fixed to a little cylinder, moving freely on a horizontal axis, which is supported by a brass pillar, secured upon one of the magnetic poles. A small mercury trough, supported between the legs of the magnet, is connected, under the stand, with one of the screw-cups. Thus the current is conveyed to the mercury. The other cup is connected with the magnet, and, through the pillar, with the wire, W.

Fig. 62.

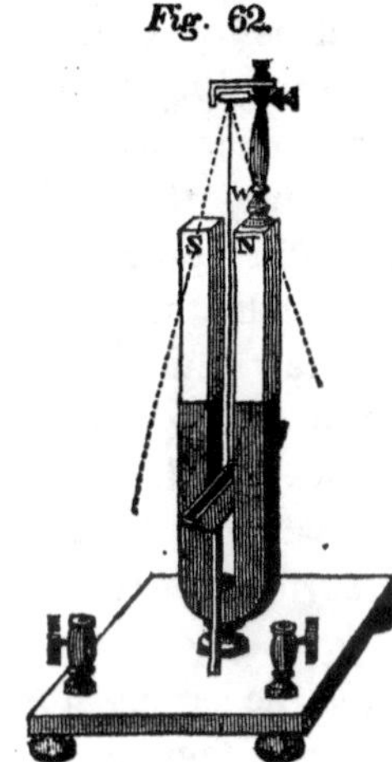

178. A single magnetic pole, placed either in a horizontal or vertical position, by the side of a wire suspended as in the last two figures, will cause it to vibrate; but its motion is less active than with two. The tendency of the wire is to revolve round the pole presented to it, as has been explained in § 154; and, when suspended between a north and south pole, simultaneously around both

179. Gold Leaf Galvanoscope.—In this instrument, represented in Fig. 63, a narrow slip of gold

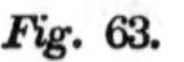
Fig. 63.

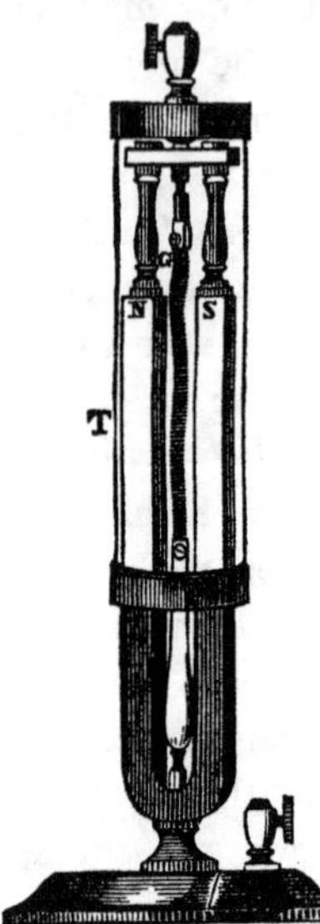

leaf, G, is suspended between the poles of a U-magnet. The legs of the magnet are brought near to each other, in order to render their action on the current conveyed by the gold leaf more powerful, and also to allow of the whole arrangement being enclosed in a wide glass tube, T. This prevents the interference of currents of air, and secures the gold leaf from injury. The upper end of the slip is in metallic connection with the screw cup surmounting the tube. Its lower end communicates with the screw cup on the stand. When a very feeble current of electricity is transmitted through the gold leaf, it becomes curved forwards or backwards, according to the course of the current; in either case, tending to move away from between the magnetic poles in a lateral direction. The motion depends on the same cause as that of the wire in the instrument described in § 177. This instrument does not indicate the quantity of the electrical current, but is an exceedingly delicate test of its existence and direction. A powerful current would of course destroy the gold leaf.

180. Revolving Spur-Wheel.—The reciprocat-

ing movement in the instrument described in § 177 may be converted into one of rotation, by making use of a copper wheel, the circumference of which is cut into rays, instead of the wire. The points of the wheel, W (Fig. 64,) dip into mercury contained in a small trough, T, supported between the legs of a U-magnet, fixed in an upright position. The axis of the wheel is supported by two brass pillars upon the poles, N and S, of the magnet. One of the screw-cups on the stand is connected with the magnet, and through that with the wheel. The mercury trough, which is insulated from the magnet, is connected with the other cup.

Fig. 64.

181. When connection is made with the battery, the current passes from the axis of the wheel to the trough through any one of the points which happens to touch the mercury. Under these circumstances, the ray through which the current is flowing passes forward between the poles of the magnet, like the vibrating wire in Fig. 62, until it rises out of the mercury. At this moment the next succeeding ray enters it, and goes through the same process; and so on.

182. If the quantity of mercury is so adjusted that one ray shall quit its surface before the succeeding one touches it, a spark will be seen at each rupture of contact. When the machine is set in motion in the dark, so that it may be illuminated by the rapid succession of these sparks, the revolving wheel will appear to be nearly at rest; exhibiting only a quick vibratory movement, in consequence of the sparks not succeeding each other precisely at the same point. This optical illusion arises from the fact, that the electric light is so extremely transient in its duration, that the wheel has not time to move to any appreciable extent during the electrical discharge; and therefore each spark shows it in an apparently stationary position. If the sparks occur at one place more frequently than at the rate of eight in a second of time, the eye cannot appreciate them separately, and the impression of a continuous light is received. For this reason, the wheel is seen constantly, as if it were illuminated by a steady light, instead of an intermitting one

183. A more rapid revolution will be obtained if a small electro-magnet be substituted for the steel magnet. The electro-magnet is included in the circuit with the spur-wheel, so that the current flows through them in succession. Here the direction of the rotation will not be changed by reversing that of the current, since the polarity of the electro-magnet is also reversed.

184. Revolving Disc.—It is not essential to divide the wheel into rays, in order to obtain rota-

tion. A circular metallic disc will revolve equally well between the poles of a magnet. In this case, the electric circuit remains uninterrupted during the entire revolution, and no sparks appear, as with the spur-wheel.

185. De la Rive's Ring. — A coil of wire, while transmitting the electric current, exhibits the same reactions with a magnet as the straight wire in Figs. 61 and 62; but the circular direction in which the current flows gives to the coil an apparent magnetic polarity. This fact may be shown by means of the apparatus figured in the adjoining cut. One end of the wire forming the coil, or helix, C, is soldered to a very small plate of copper, c, and the other to a similar plate of zinc, z. These plates are fastened to a small piece of wood, in order to keep them apart, and placed in a little glass cup, D. To put the instrument in action, a sufficient quantity of water, acidulated by a few drops of sulphuric or nitric acid, is poured into the glass cup to cover the plates, and the whole apparatus is floated in a basin of water. The coil will now be found to place itself with its axis north and south, being influenced by

Fig. 65.

the polarity of the earth, like a compass needle. The arrow indicates the course of the galvanic current in the coil from the copper to the zinc.

186. Take a bar magnet, M, and, holding it horizontally, bring its north pole near to the south pole of the ring. The ring will move towards the magnet, and pass over it until it reaches its middle, where it will rest in a state of equilibrium; returning to this position, if moved towards either pole and then left at liberty. Now, holding the ring in its position, withdraw the magnet, and pass it again half way through the coil, but with its poles reversed. The ring, when set at liberty, will, unless placed exactly at the centre, move towards the pole which is nearest, and, passing on till clear of the magnet, will turn round and present its other face. It will then be attracted, and pass again over the pole till it rests in equilibrium at the middle of the magnet.

Fig. 66.

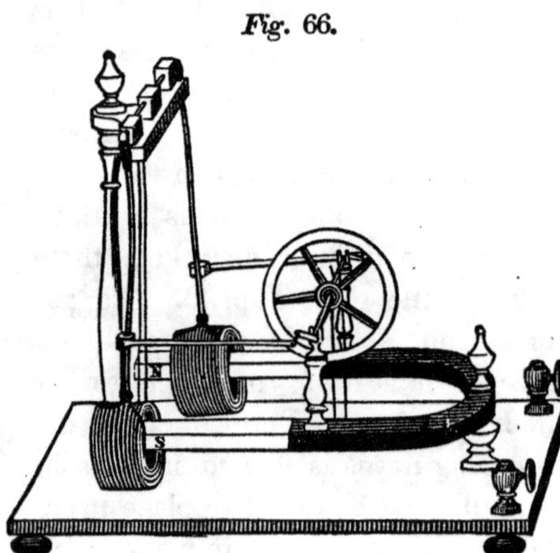

187. VIBRATING COIL ENGINE.—In the instrument represented in Fig. 66, two heliacal coils are so suspended from an arm attached to the tall brass pillar as to al-

low of their moving to some distance over the poles, N and S, of a steel magnet. When uninfluenced by the electric current, the coils hang vertically from their points of support. There are two cranks on the axis of the balance-wheel above the magnet, which convert the vibration of the coils into revolution. On opposite sides of the axis are two wires, tipped with silver springs; each of these is connected under the base-board with one end of the wire forming each coil. At the point where the springs press, a part of the axis is cut away, leaving a third or a quarter of the cylindrical surface prominent; over this is soldered a thin slip of silver. The effect of this arrangement, which is called a *break-piece*, is, that each spring alternately transmits the electric current from the axis to its coil during a part of the revolution of the wheel, while pressing upon the cylindrical portion of the axis. The screw-cups are connected, one with the axis of the wheel, the other with one end of the wire of both coils.

188. Connection being made with a galvanic battery, the current traverses one of the coils, which then moves over the pole which it surrounds, towards the middle of the magnet, in the same manner as De la Rive's ring. When it has passed some distance, the motion produced in the balance-wheel interrupts the current by breaking the contact between the cylindrical part of the axis and the spring, and the coil falls back to its first position. As the wheel moves on, the other spring comes in contact with the axis, causing the other coil to be charged, and to go through the same movements. The coils

thus vibrate alternately, producing a rapid rotation of the balance-wheel.

189. Reciprocating Coil Engine. — Fig. 67 represents an instrument in which two coils, C C, mount-

Fig. 67.

ed upon wheels, pass over the legs of a steel magnet fixed in a horizontal position, the wheels moving upon inclined rails. When the instrument is connected with the battery, the coils become charged, and move along the poles towards the bend of the magnet. The battery connections must be so made that the current shall pass in the proper direction, or the coils will tend to move away from the magnet rather than over it. When the coils reach the bend of the magnet, the platform upon which they are secured comes in contact with the wire-spring, A, and, by the movement produced in that, breaks the contact at B. The current being interrupted, the coils slide down the rails by the force of gravity, until the platform reaches a pin on the rod near B, whose movement renews the contact.

190. Reciprocating Magnet Engine. — Fig. 68

represents an instrument on the same principle as the last, except that the coils are stationary and the mag-

Fig. 68.

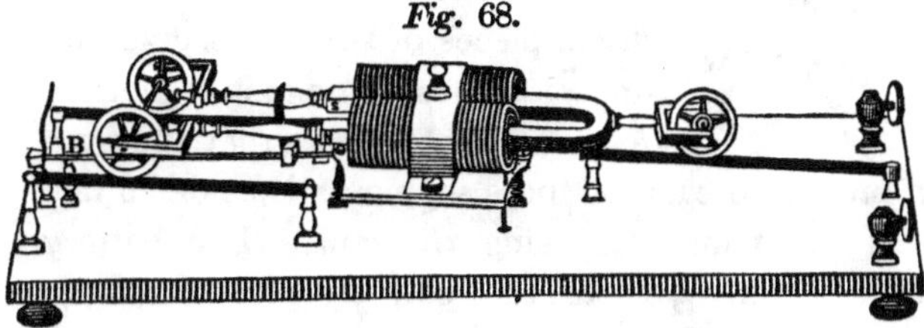

net moves within them. It is supported by wheels moving upon the inclined rails. When the bend of the magnet reaches the coils, the axle of the wheels near B strikes a spring, and breaks contact at B, by pushing along a little rod below the wheels. As the magnet falls back by gravity, the axle touches another spring on the same rod, drawing it along until a pin upon the rod strikes the pillar at B, and renews the circuit.

Fig. 69.

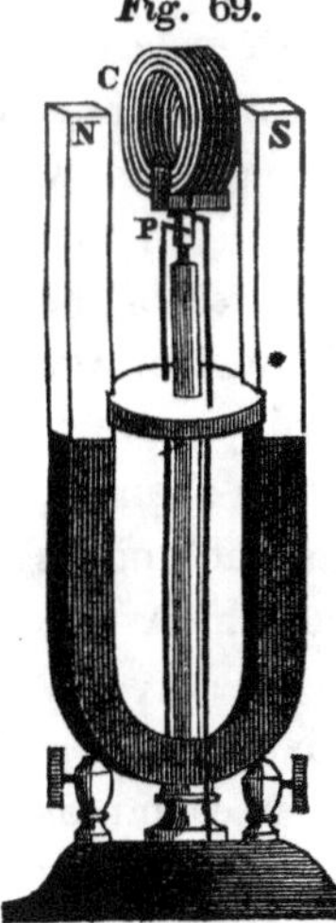

191. Revolving Coil. — This instrument consists of a U-magnet fixed upon a stand, in a vertical position, and a circular coil of insulated copper wire, C (Fig. 69), so arranged as to revolve on a vertical axis between the magnetic poles. The rotation is effected in a different manner from any previously mentioned. The polarity of the coil is reversed twice in each revolution, by means of the *pole-changer*, invented by Dr. Page, which is employed in many of the instruments to be hereafter described.

The pole-changer attached to the coil is seen at P, and a horizontal section of it is shown in Fig. 70.

Fig. 70.

It consists of two small semi-cylindrical pieces of silver, s s, fixed on opposite sides of the axis of motion, A, but insulated from that and from each other; to each of these segments is soldered one end of the wire composing the coil. The battery current is conveyed to the coil by means of two wires terminated by horizontal portions of flattened silver wire, W W, which press slightly on opposite sides of the pole-changer, and must be so arranged that the direction of the current in the coil may be reversed at the moment when its axis is passing between the poles of the magnet.

192. The coil being placed with its axis at right angles to the plane of the poles, and connection made with a battery, one extremity of the axis, or, in other words, one face of the coil, acquires north polarity, and the other south polarity, in the same manner as De la Rive's ring. The action of the magnet now causes it to move round a quarter of a circle in one direction or the other, according to the course of the current, so as to bring its poles between those of the magnet. In this position it would remain, were it not that, as soon as it reaches it, the pole-changer, which is carried round with it, presents each of its segments to that stationary silver spring which was before in contact with the opposite segment. By this movement, the current in the coil is first interrupted for a moment, and, as the coil passes on, is immediately renewed in the contrary direction, thus

reversing the polarity. Each end of the axis being now repelled by the magnetic pole which previously attracted it, the coil turns half way round, so as to present its opposite faces to the poles. At this point, the direction of the current is again reversed, causing the motion to be continued in the same direction; thus producing a rapid revolution. Instead of a coil of large diameter, the wire may be coiled into a long helix of small diameter, which will rotate in the same manner.

193. REVOLVING RECTANGLE. — This instrument is similar in principle to the last described, a rectangular coil of wire, C (Fig. 71), being substituted for the ring. The rotation is much more rapid in consequence of the proximity of the rectangle to the magnet, not only near its poles, but throughout the greater part of its length. By this means, the great speed of eight or ten thousand revolutions in a minute may be attained. In the cut, the two silver springs, which press on the pole-changer, are seen at *a*, each of them attached to a stout brass wire, proceeding from one of the cups for battery

Fig. 71.

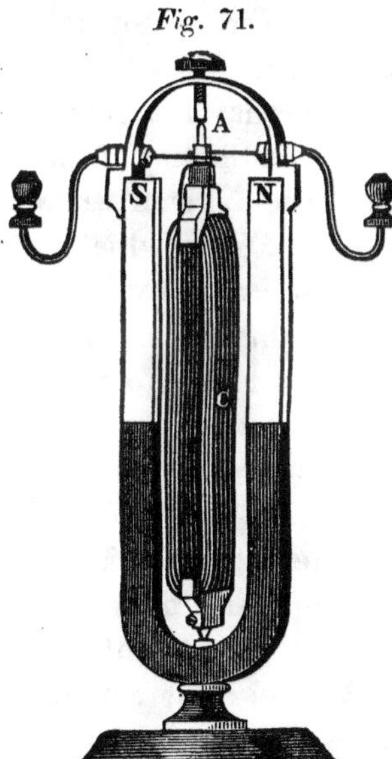

connection; these wires pass through the brass arch surmounting the U-magnet, but are insulated from it.

194. REVOLVING GALVANOMETER NEEDLE. — Fig. 72 represents an instrument of similar construction to the Upright Galvanometer (Fig. 55), except that a continued revolution of the magnetic needle is obtained by the following arrangement: One end of the wire composing the coil connects with one of the screw-cups on the stand, and its other end with the bearings of the shaft on which the needle revolves. On the shaft is a break-piece, seen at B, in the front view of the needle, given separately in the cut, upon which a wire, W, connected with the other screw-cup, presses during half of the revolution. The instrument may be used as a galvanometer, by removing W from the break-piece, and bringing it into contact with a silver pin, not shown in the cut, which connects it directly with the coil, without the intervention of the break-piece. If the direction of the battery current is reversed, the needle revolves in the opposite direction.

Fig. 72.

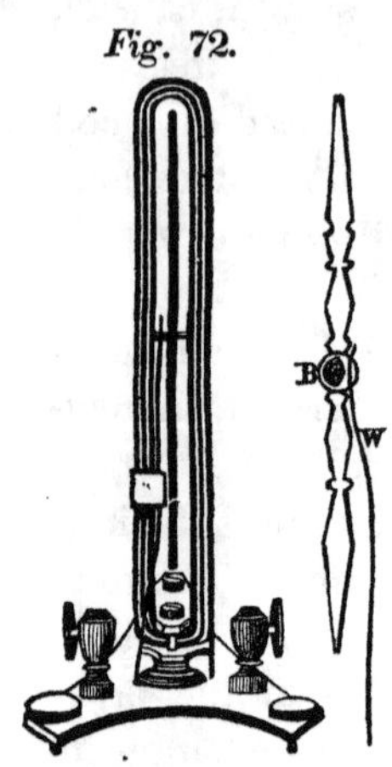

195. REVOLVING COIL AND MAGNET. — In this instrument, represented in Fig. 73, a circular coil of wire, C, is fitted to revolve, on a vertical axis, between the poles of two steel magnets. So far it resembles in principle the Revolving Coil (Fig. 69); but in

this instance the magnets rotate also. For this purpose, they are made of thin steel, and bent into a semicircular form. The two are connected by strips of brass, so as to form a circle, the similar poles being near each other, as indicated by the letters in the cut. This circle is a little larger than the coil, and revolves freely round it on a vertical axis. A peculiar arrangement is required, in order to transmit the battery current to the pole-changer belonging to the coil. The springs which press upon it are connected with two small cylinders of silver, surrounding the shaft of the magnet and insulated from it, one being a little below the other; or the cylindrical shaft itself may answer for one of them. The wires proceeding from the screw-cups on the stand press upon these cylinders. In this manner the current is conveyed to the springs bearing upon the pole-changer in a constant direction, notwithstanding that they are carried round with the magnet in its revolutions. When connection is made with the battery, the mutual action between the coil and the magnet causes them both to revolve, but in contrary directions; on the well-known mechanical principle, that action and reaction are always equal and

Fig. 73.

opposite to each other. On changing the direction of the galvanic current, both the wire coil and the magnets revolve in the opposite direction.

196. Thermo-Electric Revolving Arch. — It has been shown that when a galvanic current flows through a helix, such as De la Rive's ring, § 185, its faces acquire polarity, and, if free to move, arrange themselves north and south. In Fig. 74 there is a stand, supporting an upright brass pillar with an agate cup at the top. On this is balanced by a pivot, at A, an arch of brass wire, the two ends of which are connected by a German silver wire encircling the pillar.

Fig. 74.

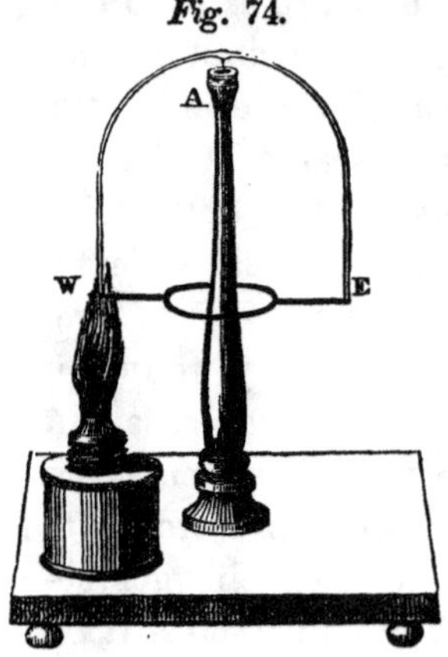

197. If the stand be arranged according to the points of the compass, and one of the junctions of the brass and German silver be heated by a spirit lamp on the east side of the stand at E, a thermo-electric current will be set in motion from the German silver, through the heated junction, to the brass, and back, through the arch, to the German silver. The current, thus established, gives polarity to the faces of the arch, as if it were a coil or helix; circulating in such a direction, that the face, which is turned towards the *north*, exhibits south polarity. Since the magnetic pole of the earth there situated is itself a

south pole, similar poles are presented towards each other, and the arch is obliged to make a semi-revolution on its axis, in order to present its northern face to this pole. This movement brings the other junction into the flame, and a current is produced opposite to the former one, which changes the polarity of the arch, and obliges it to move on through another semi-revolution. Thus the currents are reversed, and slow rotation ensues. This is probably the most delicate reaction between the magnetism of the earth and a current of electricity which has been observed.

198. If the lamp be put to the south of east, the heated junction of the arch will move round by the south; if it be put to the north of east, the heated junction will move round by the north; just as a compass needle, if its north pole is made to point south, returns to its natural position either by the east or west, if it is inclined to the one or the other. If the spirit lamp be placed exactly west, or at W in the figure, the current which is excited tends to keep the arch stationary, by causing the face which exhibits north polarity to be directed towards the *south magnetic pole* of the earth.

199. THERMO-ELECTRIC REVOLVING ARCH ON U-MAGNET.—If a thermo-electric arch A B (Fig. 75), similar to the one just described, be balanced on one of the poles of a U-magnet, the reaction between the polarity induced in it, by heating one of its junctions, and the magnetism of the opposite pole of the magnet, will be much more energetic than in the former case with the earth. It resembles, in principle, the

Revolving Coil, § 191, except that it is attracted and repelled by a single pole instead of two, the pole on which it is supported having no influence upon it. In this and other instruments of the same kind, the upper part of the arch may, with equal advantage, be of silver instead of brass.

Fig. 75.

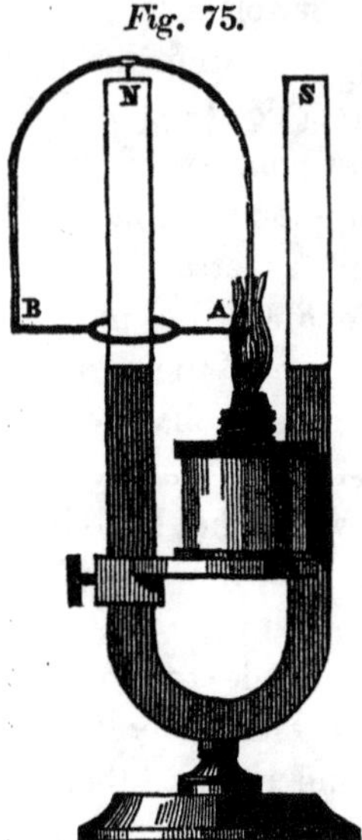

200. The most favorable position for the lamp is not that represented in the figure, but at a right angle with the line connecting the two poles, and in a line with the pole on which the frame is mounted; or in a situation analogous to the east side of the stand of the last-described instrument. By varying the lamp to one side or the other of this position, the arch will revolve in either direction, as before. On the opposite side of the pole, the lamp would have no tendency to produce revolution; though, if the arch were mounted on the south pole, the lamp should be on the farther side of the magnet, and in a line with that pole, in order to cause rotation.

201. Thermo-Electric Revolving Wire Frames. — This instrument, represented in Fig. 76, consists of two frames mounted upon the poles of a U-magnet. These frames are formed of two arches, or rather rectangles. similar in construction to that in the

instrument last described, crossing each other at right angles; and they act on the same principle as that,

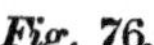

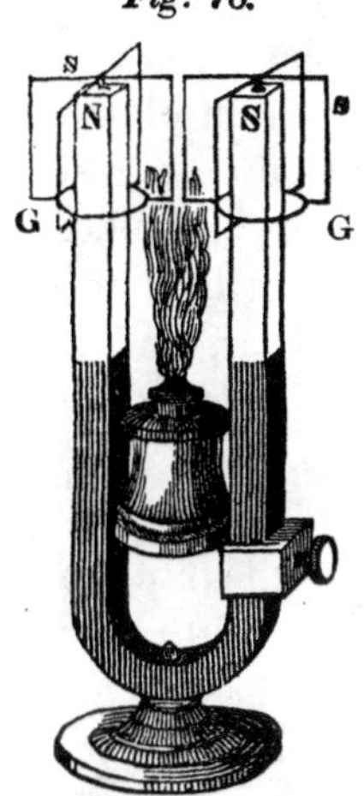

the second rectangle only contributing to the rotation produced by the first. In each individual rectangle the current is reversed every half revolution. These are generally made of silver and platinum; but German silver, in combination with brass or silver, occasions a stronger current. The lower horizontal portions of the frames, marked G G in the cut, are composed of German silver, and the other parts, s s, of silver. A frame is usually mounted on each pole; the attractions and repulsions of each frame proceeding altogether from the opposite pole. In order to heat the junctions of both frames at once, the lamp is placed between the two poles, by which there is a loss of attraction and repulsion to each frame through the distance of 90°, in which the heat would act, if two lamps were employed at right angles to the line of junction of the poles.

202. Thermo-Electric Arch rotating between the Poles of a U-Magnet. Fig. 77 represents a thermo-electric arch, mounted upon a brass pillar, between the poles of a horseshoe magnet; the circular part is of German silver, and the upper part of silver. In this case, both poles conspire in producing revolution, the motion of the arch depending upon

the same principle as that of the Revolving Coil. Yet the different mode of reversing the current in this instrument occasions the arch to rotate in either direction when the lamp is in front of the magnet, and to remain at rest when the lamp is on the other side. A stand to support the lamp slides on the brass pillar, and is fixed at any required height by means of a binding screw. The lamp should be placed in the position represented in the cut, in front of the magnet, its north pole being on the left.

Fig. 77.

203. When either of the junctions is in the flame, a current flows from the German silver to the silver, ascending by the heated side of the arch, and descending by the other. That face which is presented towards the north pole acquires north polarity, and the other face south polarity. The influence of the magnet now causes the arch to turn half way round, so as to present its southern face to the north pole. This movement brings the other junction into the flame; the polarity of the arch is reversed, and it moves on as before.

204. If the lamp be placed in the corresponding position on the other side of the magnet, the direction of the current will be such that the southern

face of the arch will be presented towards the north pole. In this position the arch tends to remain, returning to it when moved to either side; and consequently no revolution can be obtained. Care should be taken not to allow the junction to remain so long in the flame as to melt the solder.

205. Double Thermo-Electric Revolving Arch. — In this instrument, two arches, *a* and *b* (Fig. 78), are so mounted as to revolve between the poles of a U-magnet fixed in a horizontal position. The horizontal portion of the arch *a* is of German silver, and the upper part of silver; while in *b*, the lower portion is of silver, and the upper part of German silver. A single lamp is so placed as to heat both arches; the current excited in each ascends on its right side and descends on its left side, because the heat is applied to the right junction of *a* and to the left of *b*. Each of them now presents a north pole towards the north pole of the magnet, the currents circulating in the opposite direction to that of the hands of a watch. They consequently both revolve, either in the same or in opposite directions. If the arches be transposed, so that *b* occupies the place of *a*, neither of them will move so long as the lamp is in the position represented in the cut.

Fig. 78.

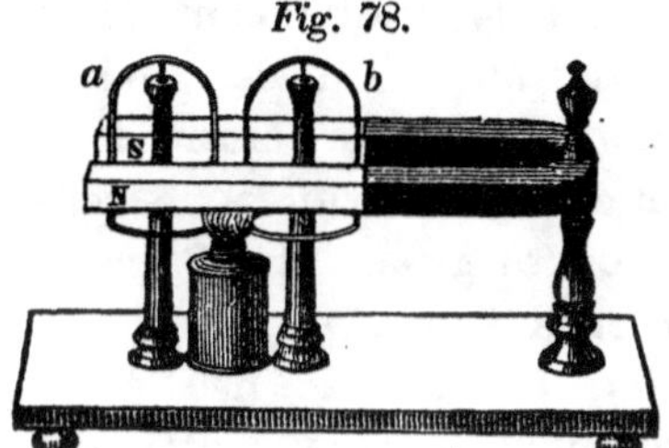

III. IN REFERENCE TO THE EARTH.

206. A magnetic needle so suspended as to move only horizontally, assumes, when left to itself, a determinate direction with respect to the earth. Its poles place themselves nearly in a plane passing through the geographical poles of the earth; this direction consequently corresponds nearly with the meridian of the place where the observation is made. When displaced from this position, the needle returns to it after a series of oscillations. This directive tendency, and its probable discovery by the Chinese, have already been mentioned in § 11.

207. Fig. 79 represents a magnetic needle, so arranged as to move freely in a horizontal direction only. The mode of suspension is the same as in the compass needle; the latter, however, is fixed to a circular card which of course moves with it, and on which the cardinal points are marked.

Fig. 79.

208. If the needle be suspended so as to have freedom of motion in a vertical direction, it is found not to maintain a horizontal position, but one of its poles (in this hemisphere the north) inclines downwards towards

the earth. At the magnetic poles of the earth, the direction of the needle would be vertical; but the inclination diminishes as we recede from the poles towards the equator, and at the magnetic equator, which does not vary greatly from the geographical one, the needle becomes horizontal. A needle properly prepared for exhibiting this inclination, is called a Dipping Needle.

209. Fig. 80 represents a dipping needle whose mode of suspension allows of its turning freely in any direction. It is fixed, by means of a *universal joint*, to a brass cap containing an agate, which rests upon the pivot. The usual arrangement allows only of motion in a vertical plane, the needle having an axis passing through its middle at right angles to its length, which axis is supported horizontally. This form is represented in Fig. 6. The small needles in Fig. 81 are mounted in the same manner. Sometimes a vertical graduated circle is added, to measure the angle which the needle makes with the horizon. In using a needle whose motion is confined to a single plane, it must be so placed that this plane may be directed north and south, coinciding with the plane of the magnetic meridian. A dipping needle, before being magnetized, should be as equally balanced as possible, so as to emain at

Fig. 80.

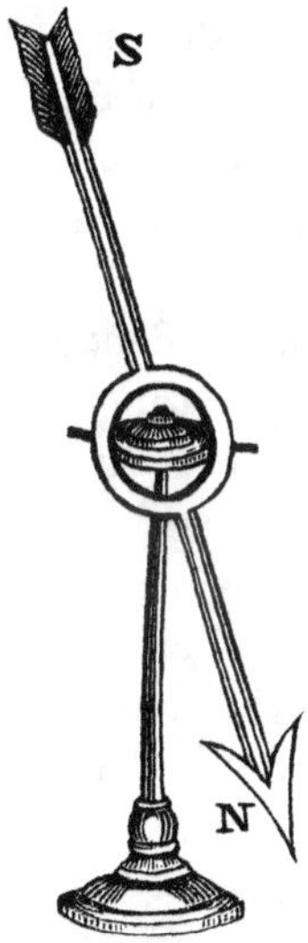

S

N

rest in any direction in which it may be placed; a high degree of accuracy is, however, difficult of attainment.

210. The dipping needle assumes, in various latitudes, the directions exhibited in the annexed diagram (Fig. 81), where the point of the arrow indicates the north pole, and the feather the south pole of the needles placed around the globe. The angle which the needle makes with the horizon at any place is called the *dip*, at that place. The tendency of the needle to dip is counteracted in the mariner's and surveyor's compasses, by making the south ends of needles intended to be used in northern latitudes somewhat heavier than the north ends. A needle well balanced in this latitude remains sufficiently horizontal for use in any part of the northern temperate zone, except for delicate experiments. Compass needles intended to be employed on voyages or travels, where great variations in latitude may be expected, are provided with a small weight sliding upon the magnet. By a proper adjustment of this, the greater or less tendency to dip is counterbalanced. In passing from the northern to the southern hemisphere, the weight must be transferred from the south to the north pole of the needle.

Fig. 81.

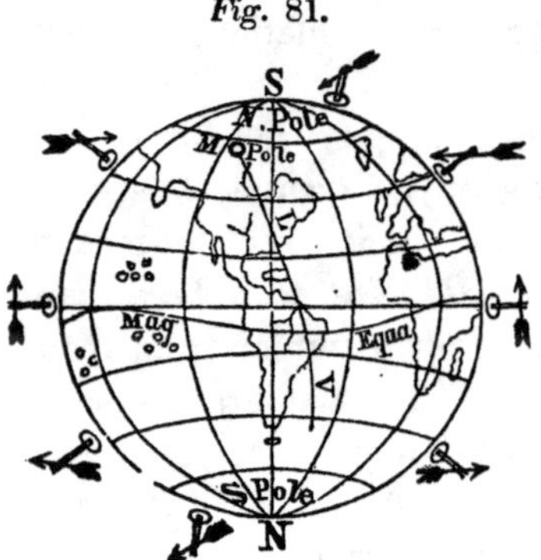

211. In Fig. 81, the North American magnetic pole is represented near S, the north geographical pole of the earth. The line L V is nearly the present *line of no variation* (see § 215), and the curved line near the geographical equator is the magnetic equator, where the dip is at zero, and the direction of the dipping needle the same as that of the horizontal needle.

212. By comparing the directions assumed by the needle in its various positions with regard to the earth, as represented in Fig. 81, with those assumed by a magnet in reference to another magnet, as illustrated in Fig. 47, a great similarity is found to exist between them. This led to the opinion, which was for a long time entertained, that the earth was itself a magnet, or that it contained within it large magnetic bodies, under the influence of which the magnetic needle assumed these various directions; as a small needle does when placed in various positions near to a bar magnet.

213. But there is another mode of accounting for the directive tendency of the magnet in respect to the earth; and that is, by supposing, instead of magnetized bodies within the earth, lying parallel to the direction of the needle, currents of electricity passing around the earth, within it, but near the surface, at right angles with that direction. This would identify the directive power of the needle in respect to the earth, with its directive tendency in regard to a current of electricity, as described under the last head. And this is, in fact, the view at present adopted.

214. The direction of the needle in respect to the earth is not fixed. Its *variation*, that is, its deviation from the true geographical meridian, is subject to several changes, more or less regular. So also is the intensity of the action exerted on it by the earth, as shown by the number of oscillations made by it in a given time. When examined by means of apparatus constructed with great delicacy, the needle is found to be seldom at rest.

215. The instrument represented in Fig. 82 is intended to illustrate the magnetism of the earth, on the supposition stated in § 212. The compound bar magnet, *n s*, is placed in the magnetic axis of the earth, not coinciding exactly with the axis of rotation, N. S. A small magnetic needle, placed at B, on the magnetic meridian, will point both to the magnetic pole *s*, and to the north pole, N, both being in the same line. But, if the needle be placed at A, or any where except on the magnetic meridian, it will point to the magnetic pole alone, the two poles not being in the same direction. The several magnets represented at *n s* are not fastened together, but only fixed on one axis. This allows their poles to be separated a little, to imitate more closely the distribution of terrestrial magnetism; the

Fig. 82.

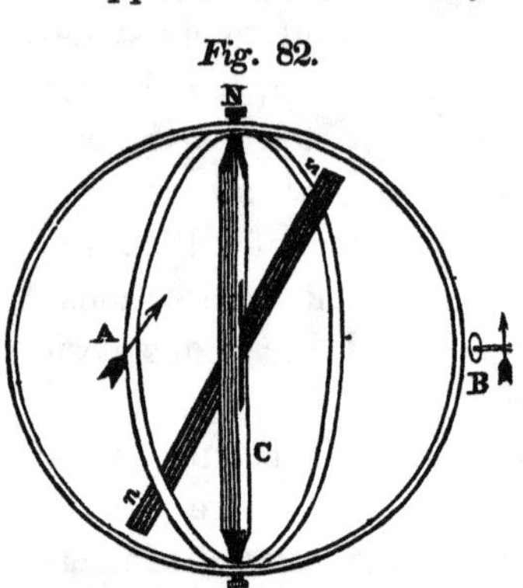

earth really having four magnetic poles, two strong and two weak: the strongest north pole is in America, the weakest in Asia. The line of no variation on the earth's surface differs considerably from the magnetic meridian, and the lines of equal variation and equal dip are not exactly meridians and parallels of latitude to the magnetic pole. The action of the magnetism of the earth at its surface is, therefore, irregular.

216. The variation of the needle at any place is found by observing the magnetic bearing of any heavenly body whose true position at the time is known. It is immediately obtained by comparing the direction of the needle with the north star when it crosses the meridian, or by calculation when the north star is at its greatest elongation. An observation of the sun, however, is usually preferred. The latitude of a place, A (Fig. 83), being known, the exact bearing of the sun, S, east or west, can be obtained by calculation,* for any given moment of time at that place. If the needle at A points to M, instead of N, the true north, the

Fig. 83.

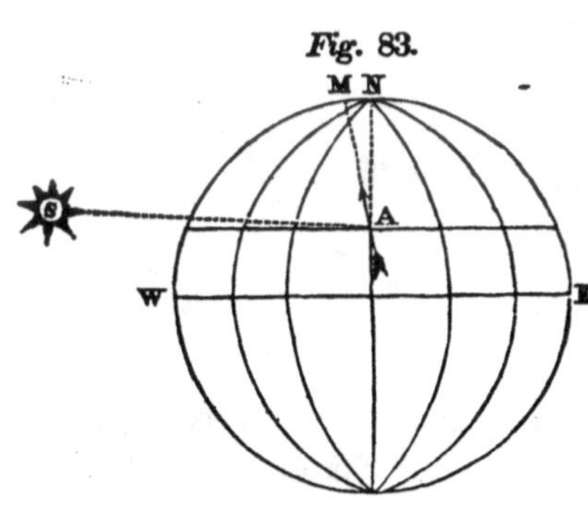

* See Bowditch's Navigator.

angle M A S will be the magnetic bearing of the sun west. Suppose this angle to be observed by the surveyor's compass, and found equal to 76°, the time being exactly noted. The angle N A S, the true bearing of the sun at the time, is then calculated. Suppose it equal to 85° 30′. The difference between the magnetic bearing and the true bearing, represented by the angle M A N, is the variation of the needle, and equals 9° 30′.*

217. Fig. 84 represents an instrument intended to illustrate the theory which ascribes the magnetism of the earth to electrical currents circulating around it at right angles to its axis. N S is merely a wooden axis to the globe. When a galvanic current is sent

Fig. 84.

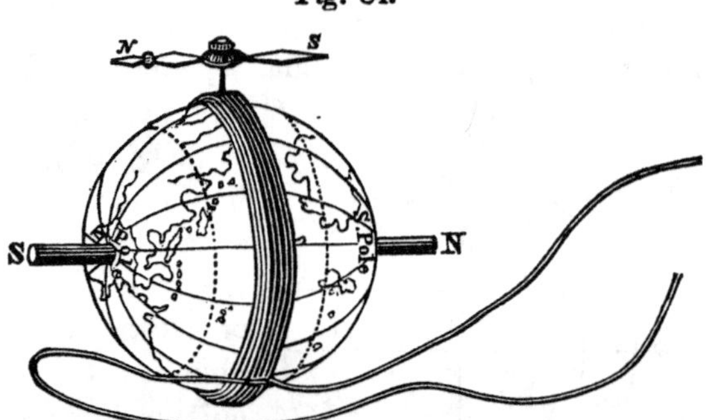

through the coil of wire about the equatorial region, small magnetic needles, placed in different situations, arrange themselves as they would in similar terrestrial latitudes. By comparing this figure with Fig.

* The variation at Boston, in May, 1847, was 9 deg. 30 min. west. The dip, at the same time, was 74 deg. 20 min.

82, representing the globe with the included magnet, a comparison may be made between the two theories of magnetism. The needles arrange themselves similarly on both globes. With a small dipping needle, the resemblance between its positions on either and those assumed by it on the earth's surface is very striking.

218. It will be observed that, in Fig. 82, the *south* pole of the included magnet is represented near the *north* geographical pole of the earth. So, also, in Fig. 84, the rod, N S, passed through the axis of the globe, shows the direction of the polarity induced by the current to be contrary to that of the geographical poles. The reason of this may be easily understood. The northern magnetic pole is the one which *attracts* the north pole of a magnet, and therefore must itself possess south polarity, and not north, as its name might seem to indicate. In Fig. 84, the battery current is considered as flowing round the globe in the same direction as the supposed currents in the earth; that is to say, from east to west, in the opposite direction to that of the earth's rotation. The apparent magnetic polarity induced in a coil of wire by the electric current, which occasions the needle to assume these directions, will be described under a subsequent head.

219. The aurora borealis is found to affect a delicately-suspended magnetic needle, causing it to vibrate constantly, but irregularly, during its continuance, and especially when the auroral beams rise to the zenith; if the aurora is near the horizon, the

disturbance of the needle is very slight. When the beams unite to form a corona, its centre is generally in or near the magnetic meridian.

220. Within a few years, a considerable number of magnetic observatories have been established in various parts of the world, for the purpose of making systematic and corresponding observations in relation to terrestrial magnetism. At these stations the variation of the needle, and the intensity of the earth's action upon it, are observed and recorded almost hourly, and on stated days at intervals of a few minutes only. These observations, made by means of excellent instruments, and at the same time in widely remote regions, admit of comparison with each other, and can hardly fail to throw light on many parts of this important and intricate subject.

MAGNETISM.

II.

INDUCTION OF MAGNETISM.

I. BY THE INFLUENCE OF A MAGNET.

221. A piece of iron, of any form, brought near to a steel magnet, becomes itself magnetic by induction. Thus, let M (Fig. 85) be a bar magnet of steel, and B an iron rod brought near to it. By the influence of the magnet the rod will become magnetized; the end nearest the south pole will become north, and the end remote from it, south. The magnetic induction is stronger when the bar is brought in contact with the pole of the magnet; a decided effect, however, is produced by the mere proximity of the magnet to the iron. That the iron bar, while under the influence of the magnet, actually possesses magnetic properties, may be shown by presenting to it some iron filings or small nails, which will adhere to each extremity; and also by bringing near to it a small

Fig. 85.

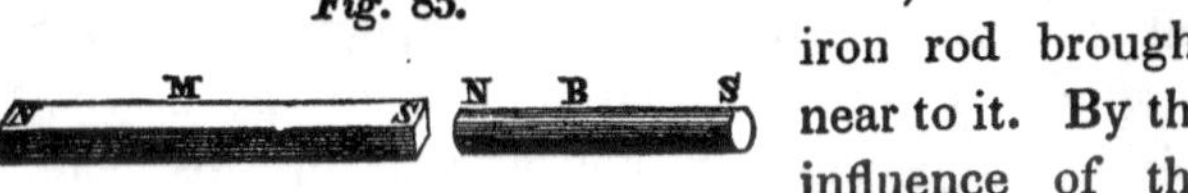

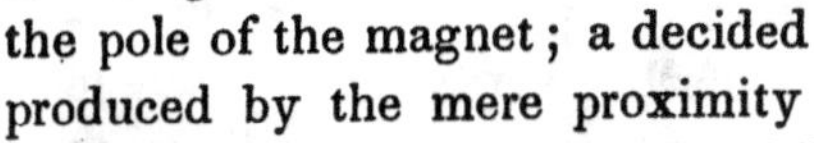

magnetic needle balanced on a pivot, the north pole of which will be repelled by the end of the bar nearest to the magnet, M, and attracted by the end farthest from M. This induced magnetism immediately disappears when the iron is removed from the vicinity of the magnet. If a small bar of steel be substituted for the iron bar, it acquires magnetism much less readily, but retains it after removal; becoming, in fact, a permanent magnet.

222. It was formerly supposed that the attractive force of the loadstone, or any other magnet, was exerted upon iron simply as iron; but it is now known to be the attraction of one pole of a magnet for the opposite pole of another magnet. In all cases, when a magnet is brought near to or in contact with any magnetizable bodies, as pieces of iron, iron filings, or ferruginous sand, all such bodies, whether large or small, become magnetized; the part which is nearest to either pole of the magnet acquiring a polarity opposite to that of this pole, while the remote extremity becomes a pole of the same name.

223. The north pole of a bar magnet, M (Fig. 86), placed on the centre of a circular plate of iron, induces south polarity in the part immediately beneath it, and a weak north polarity in the whole circumference, so that it will sustain iron filings, as shown in the cut. If an iron plate is cut into the form of a star, as in Fig. 87, each point acquires a stronger north polarity than the edge of the round plate in Fig. 86, and is able to lift several iron screws

or nails; the letters in the cut indicate the position of the poles.

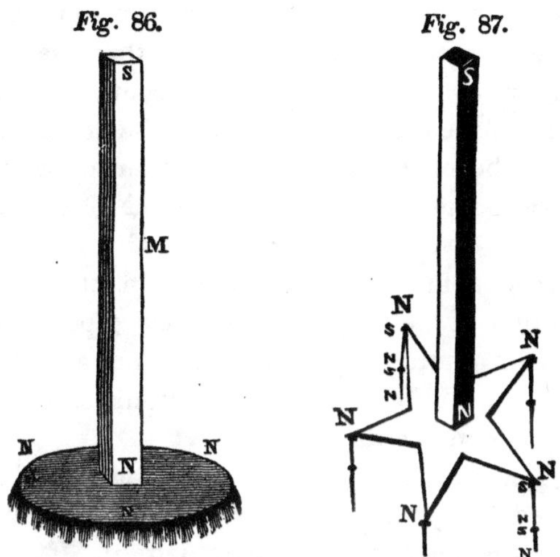

Fig. 86. *Fig.* 87.

224. Let the north pole of a steel magnet, N S (Fig. 88), be placed on the middle of a bar of iron. The part of the bar in contact with the magnet becomes a south pole, while north polarity is distributed along both halves of the bar.

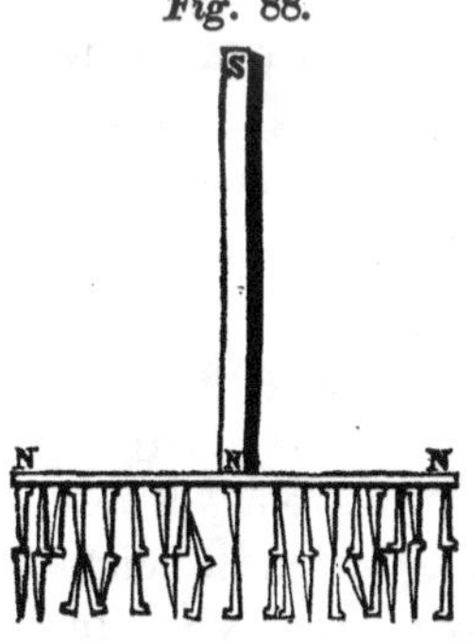

Fig. 88.

225. If several pieces of iron wire, of the same length, be suspended from a magnetic pole, they do not hang parallel; but the lower ends diverge from each other, in consequence of their all receiving the same polarity by induction.

while the upper ends are retained in their places by the attraction of the magnet. Let two short pieces of iron wire be suspended by threads of equal length, fastened to one end of each piece, so that the wires may hang in contact. If, now, the south pole of a magnet be placed below the wires, the lower ends of both will become north poles, and their upper ends south poles; and the wires will recede from each other. This divergence increases as the magnet is brought nearer, until it reaches a certain limit, when the attraction of the magnet for the lower poles overpowers their mutual repulsion, and causes them to approach each other; while the repulsion of the upper ends remains as before.

226. The mutual repulsion of two pieces of iron magnetized by induction may also be shown by the arrangement represented in Fig. 89. There are two short pieces of iron wire, upon the ends of which are secured little brass rings, which answer the purpose of rollers. The two wires being placed together, as represented by the dotted lines at A, let a U-magnet, held in a vertical position over them, be gradually brought near, keeping the poles in the direction of the length of the wires.

Fig. 89.

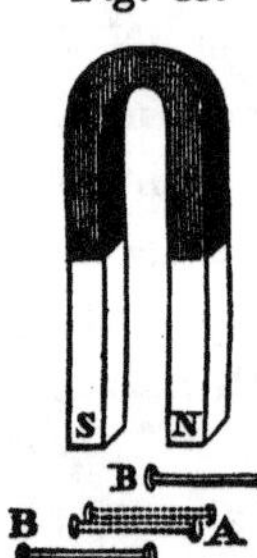

As the magnet approaches, the wires become charged, the similar poles lying in the same direction. When the magnet is brought so near as to occasion sufficient repulsion between the induced poles, the wires recede from each other, as shown at B B, in the cut.

227. Y-Armature. — This consists of a piece of soft iron, in the shape of the letter Y. If one of the branches of the fork be applied to the north pole of a horseshoe magnet, as seen in Fig. 90, the lower end of the armature, and also the other branch of the fork, acquire north polarity, and will sustain small pieces of iron. If both branches of the fork be applied, one to each pole of the magnet, as shown by the dotted lines in the cut, the polarity of the lower end immediately disappears. This is because the two poles tend to induce opposite polarities of equal intensity in the extremity of the armature, which of course neutralize each other. If the branches of the fork are applied to the *similar* poles of two magnets, their influence will conspire in inducing the same polarity in the lower end, and a greater weight will be supported by it, than when one branch is applied to a single pole.

Fig. 90.

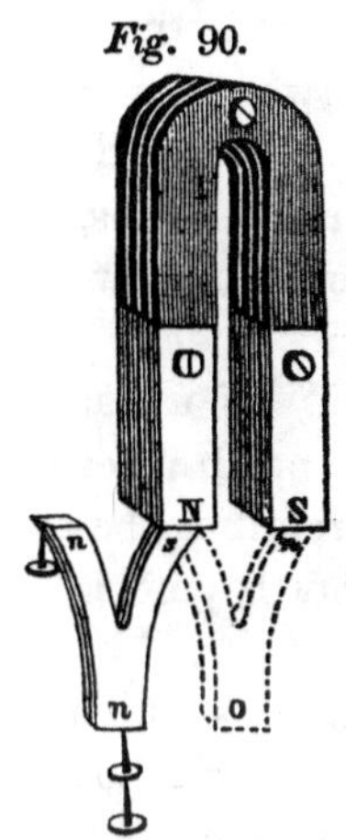

228. When the two extremities of an iron bar are acted upon at once by both poles of a magnet, there is a double induction, and the effect is greatly increased. Thus, let N S (Fig. 91) be a compound U-magnet, and A an iron armature, of such a length that, while one end is applied to the north pole of the magnet, the other extremity may be applied to the south. In this case, both poles of the magnet act, each inducing a polarity opposite to its own in that

extremity of the armature which is under its influence. The force with which the armature adheres is consequently greatly increased, there being a strong attraction between each pole of the magnet and the corresponding extremity of the armature, that is, corresponding in position; for the polarity of the parts in contact will evidently be of opposite denominations. If a bar of iron is placed between the north poles of two magnets, both extremities become south poles, while a north pole is developed at the middle of the bar.

Fig. 91.

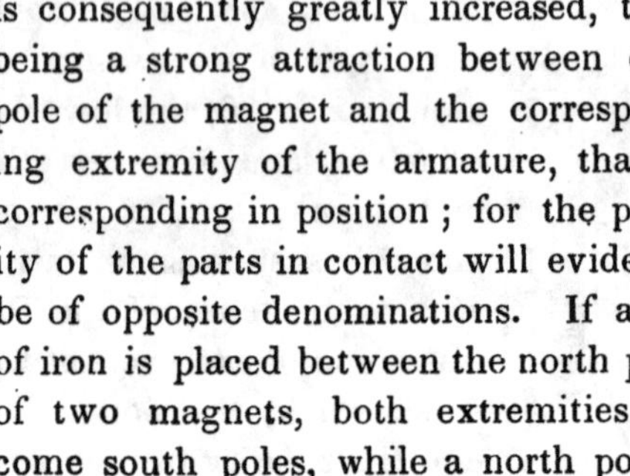

229. Rolling Armature.—This apparatus consists of a horseshoe magnet and an iron wire or rod, whose length is a little greater than the breadth of the magnet. To the middle of the wire a small fly-wheel is attached. This armature is placed across the magnet, at some distance from the poles, and the magnet held in such a position, with the poles downward, that the armature may roll towards them. When it reaches the poles, the magnetic attraction of the iron axis prevents its falling off, while the momentum acquired by the fly-wheel carries it forward, causing it to roll some distance up the other side of the magnet.

Fig. 92.

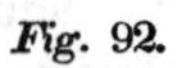

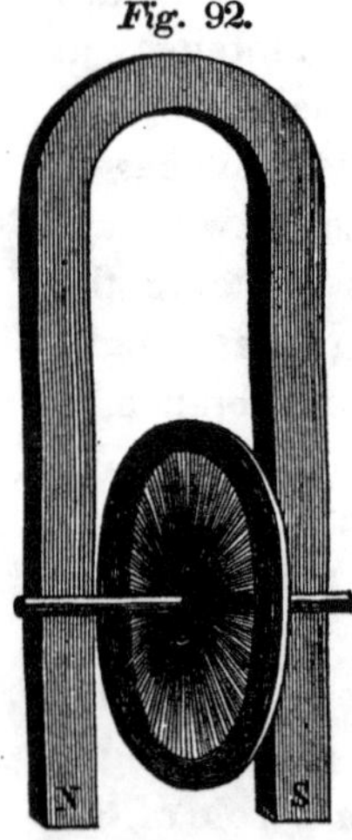

230. MAGNET AND ROLLER. — Some of the phenomena of magnetic attraction and repulsion may be shown by the arrangement represented in Fig. 93. The poles of the steel magnet, N S, are raised by a little wooden block, and a piece of iron, A, is placed across, near the bend. A short iron wire, K, has a brass ring encircling each end, which prevents actual contact between the iron and the magnet, and thus allows the wire to roll freely. When K is placed in contact with A, it immediately rolls up the magnet, in opposition to gravity, until it nearly reaches the poles. The actual poles or points of greatest magnetic intensity, in a steel magnet, are not exactly at the ends, but a little within them. They are so near that it is usual to give the name of poles to the extremities themselves. Upon bringing up the armature, R, to the poles, the wire, K, is repelled by it (see § 226), and from this cause, and the diminished action of the magnet upon it, the wire rolls back towards A. Being repelled by A, it takes a position of equilibrium near it; and when R is removed, again rolls up the magnet.

Fig. 93.

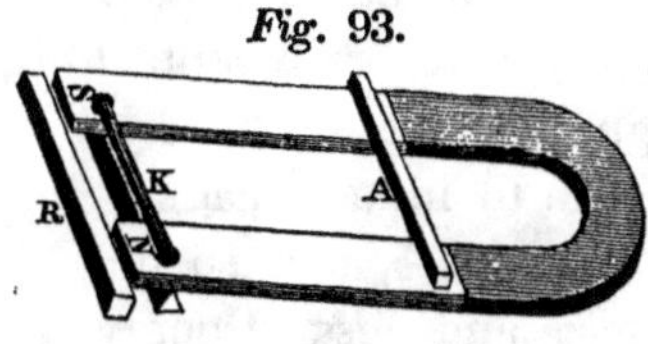

Fig. 94.

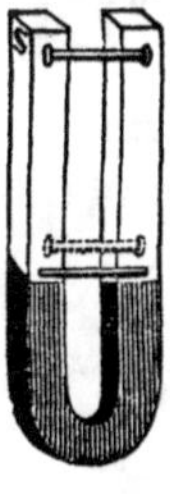

231. With a small magnet, whose poles are brought near together, as represented in Fig. 94, the attractive force is sufficient to raise the wire up to the poles, even when the magnet is placed in a vertical position.

232. Fig. 95 illustrates the successive development of magnetism in several bars of iron. The

M *Fig.* 95. *a* *b* *c*

bar *a* being placed near to or in contact with the north pole of a magnet, M, becomes itself temporarily magnetic, and is able to induce magnetism in a second bar, *b*, this again in *c*, and so on, each succeeding bar being less and less strongly magnetized. The same thing occurs with the iron nails represented in Fig. 87, hanging from the points of the star. If the magnet, M, be removed from the bar *a*, the magnetism of the whole series disappears.

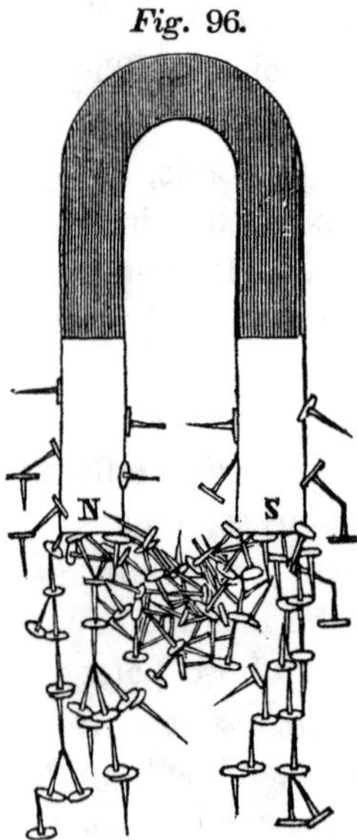

Fig. 96.

233. When a bar magnet is placed in a heap of iron filings, and then lifted up, the filings attach themselves to the poles in clusters. Each particle becoming a little magnet, they adhere together in lines, whose length depends upon the power of the steel magnet. A U-magnet, plunged into a mass of tacks, sustains long lines of them, as represented in Fig. 96. The chains of tacks suspended from the two poles attract each other, and, becoming connected, form curved lines, rudely representing the magnetic curves spoken of in § 147.

234. Loadstone. — The arrangement of small brads around the poles of a loadstone, is represented in Fig. 97. In natural magnets the poles are not usually so distinctly marked as in artificial ones, the magnetic forces being more irregularly distributed.

Fig. 97.

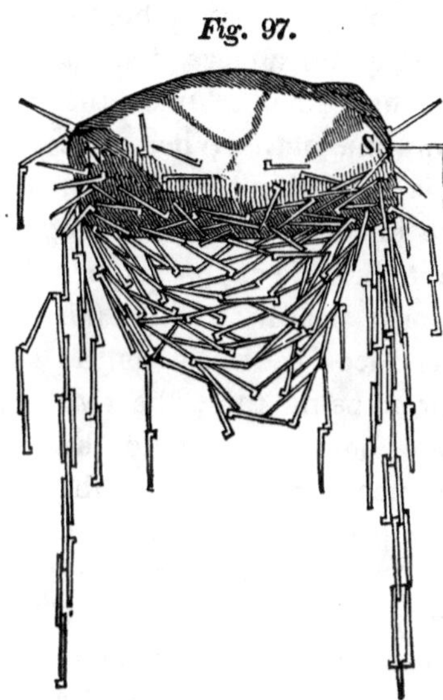

235. The attraction of a magnetic pole for a piece of iron is not exerted upon the whole mass, since both a north and a south pole are necessarily developed in the iron. Hence only that part of the piece which is nearest the magnet can be attracted, while the remote end must be repelled. If the piece of iron has any considerable length, the end which is repelled will be at such a distance from the influence of the magnet, that its repulsion will be overpowered by the attraction of the extremity which is near it. If, however, the piece is very short, so that the repelled pole is brought near to the magnet, the repulsion is proportionally stronger, neutralizing the attraction to a considerable extent; and, finally, if the iron is of such a form as to bring the two opposite poles as near together as possible, so as to expose

them nearly equally to the influence of the magnet, the attraction becomes scarcely perceptible. Thus, let M (Fig. 98) be the south pole of a bar or horseshoe magnet, and A a piece of sheet iron somewhat smaller than the end of the magnet. When this iron plate is placed in the position represented in the upper figure, the surface next the pole of the magnet acquires north polarity, while the opposite surface becomes south, and, the iron being thin, the two surfaces are both so near to the magnet that one is repelled nearly as much as the other is attracted. The thin plate will be found to adhere to the pole with a very slight force, and will tend to slip down into the position represented in the lower figure. In this position it is much more strongly attracted; for the two *ends*, instead of the two *surfaces*, become the poles, and the end in contact is attracted, and the remote end repelled. The same effect will be produced if the plate is applied to the pole of the magnet by its edge, instead of by one of its surfaces; by this means, the repelled pole of the plate is removed to a distance from the magnet.

Fig. 98.

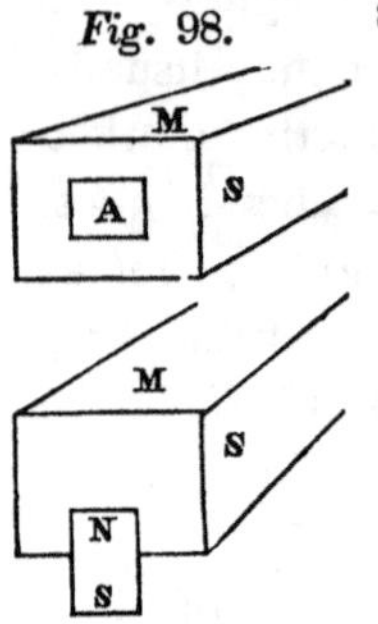

236. The inductive action of a magnet is not impeded by the interposition of any unmagnetizable body whatever. If a plate of glass be placed between the magnet and a piece of iron, the iron is attracted as strongly as it would be at the same

distance with no glass interposed. Fig. 99 represents a piece of iron wire, sealed up hermetically in a glass tube. When this is applied to a U-magnet, the iron is attracted with sufficient force to hold the tube in contact with the poles. The polarity induced in the iron is precisely as great as when only air is interposed. The result is the same when the iron is enclosed in a tube of brass, or any metal not susceptible of magnetism.

Fig. 99.

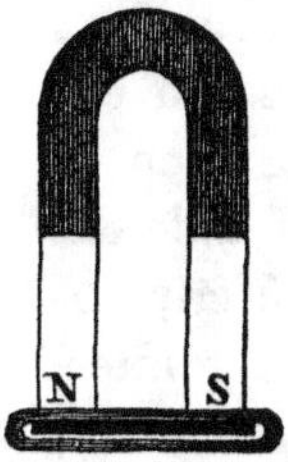

237. Fig. 100 represents an instrument designed to measure the force of attraction between a U-magnet and its armature at different distances. It

Fig. 100.

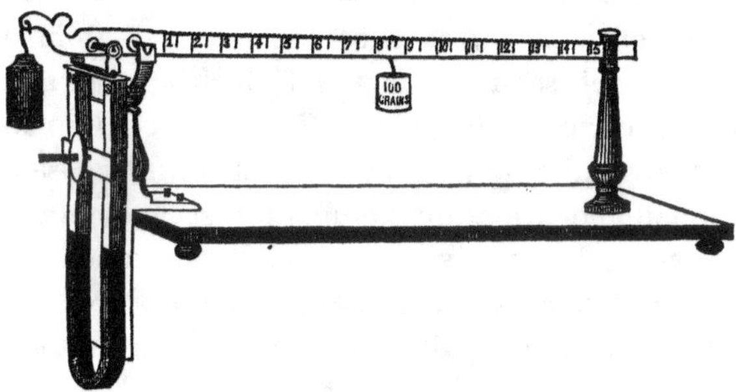

is on the principle of the steelyard. The beam, which is of brass, is supported by steel knife edges, resting in curved supports of the same metal. To the shorter arm of the beam is secured the armature, consisting of a small cylindrical bar of iron. The force with which this is held down is measured by weights hung upon the graduated longer arm. The magnet is held firmly against a vertical brass plate, by means of a clamp of the same metal, which allows the poles to be fixed at any required distance from the armature. The distance was determined in the following way: Sheet brass of $\frac{1}{240}$ of an inch in thickness was taken, and several series of different thickness formed by pressing strips of it firmly together and soldering them at the edges. By interposing these strips between the magnet and armature, they were maintained at any desired distance, from a single thickness, or $\frac{1}{240}$ of an inch, up to ten thicknesses, or $\frac{1}{24}$ of an inch

238. The following tables were constructed from experiments with two steel magnets, of different lengths, and a long electro-magnet. The first magnet employed was a very short one, formed from a steel bar, ½ inch square. It was 1¾ inches long, and 1½ inches between the poles. The first column in the table gives the number of brass strips interposed, the magnet and armature being at first in contact. The second column gives the weight sustained, or the force of attraction, in grains; and the third, the product obtained by multiplying the weight in each case by the number of interposed strips. There is a very slight spring in the brass, which diminishes as the number of strips is increased. In consequence of this, the armature is removed slightly beyond the calculated distance, and the weights in the first part of the table fall a little below the true proportion.

TABLE I.

Distance.	*Weight in Grains.*	*Product.*
0	21,200	
1	4,000	4,000
2	2,170	4,340
3	1,500	4,500
4	1,220	4,880
5	970	4,850
6	810	4,860
7	670	4,620
8	595	4,760
9	500	4,500
10	445	4,450

239. The second steel magnet was of the same thickness and width, but 11 inches long. The results obtained with this are given in Table II.

TABLE II.

Distance.	*Weight in Grains.*	*Product*
0	62,500	
1	32,000	32,000
2	17,300	34,600
3	13,100	39,300
4	9,600	38,400
5	7,800	39,000
6	6,800	40,800
7	5,600	39,200
8	4,700	37,600
9	4,050	36,450
10	3,600	36,000

240. The third series of experiments was made with an electro-magnet, $9\frac{1}{2}$ inches long, formed from a round bar, 1 inch thick. Its poles were $1\frac{1}{4}$ inches apart. The results, with one pair of Smee's battery, are given in the following table:—

TABLE III.

Distance.	*Weight in Grains.*	*Product.*
0	82,000	
1	35,000	35,000
2	25,000	50,000
3	20,000	60,000
4	15,500	62,000
5	12,100	60,500
6	11,300	67,800
7	9,300	65,100
8	7,400	59,200
9	6,500	58,500
10	5,500	55,000

241. It will be seen that the numbers in the third column of Tables I. and II. agree sufficiently well to give the following practical rule for finding the relative attractive force at different distances; that, with steel magnets of any length, the force with which the armature is attracted varies in the inverse ratio of the distance. The third column in Table III. shows that the electro-magnet does not follow this proportion. If the points of strongest polarity are regarded as situated a little within the ends of the magnet, for instance, at $\frac{1}{48}$ of an inch, or the thick-

ness of five strips of brass, in the above tables, the distances calculated from these will give very nearly the true law, as stated in § 142. A still closer approximation might probably be made.

242. Table IV. gives the comparative attractive force of the long steel magnet, with the armature in contact, and when separated by a single thickness of brass, the charge of the magnet being progressively reduced. In the first column are given the weights sustained in grains, the armature being in contact with the poles; in the second, the corresponding weights when their distance was $\frac{1}{240}$ of an inch.

TABLE IV.

In Contact.	*Distance,* $\frac{1}{240}$
62,500	32,000
52,500	28,000
33,500	10,900
6,600	2,280

243. The same instrument was used to measure the force of attraction between the opposite poles of two bar magnets. These were made from the same square bar of steel, and were each 6 inches long and $\frac{5}{16}$ of an inch thick. One of them being suspended vertically from the shorter arm of the brass beam, the other was secured to the brass plate perpendicularly below the first. The poles were placed in contact in the first experiment, and afterwards separated to different distances from $\frac{1}{240}$ to $\frac{1}{24}$ of an inch, by strips of brass, the lower magnet being moved

sufficiently to keep the beam horizontal. Beyond $\frac{1}{24}$ of an inch, the instrument was not delicate enough to give the weight with sufficient accuracy.

244. In Table V. are given the results of this series of experiments.

TABLE V.

Distance.	*Weight in Grains*
0	6,000
1	3,100
2	2,200
3	1,900
4	1,600
5	1,300
6	1,200
7	1,100
8	980
9	890
10	770

In the first experiment in the table, the poles touched by their whole surface. By moving the lower magnet laterally until the edges only of the poles were in contact, the force of attraction was considerably increased, being equal to 10,000 grs.

245. By considering the points of strongest polarity, or the true poles, as situated at a distance of four and a half strips of brass within the ends of each magnet, the results given in Table V. are made to agree closely with the law of the inverse ratio of the square of the distance.

246. The following table gives the attractive force between one of the bar magnets and a straight bar of iron, 6 inches long : —

TABLE VI.

Distance.	*Weight in Grains.*
0	5,600
1	1,480
2	870
3	670
4	520
5	410
6	360
7	300
8	270
9	240
10	200

These numbers may be brought under the law stated in § 142, by calculating the distances from a point $\frac{1}{48}$ of an inch within the steel magnet. In the iron bar, as in the armatures of the U-magnets, the poles are probably situated at the surface.

247. A close analogy exists between the phenomena of magnetism and of statical electricity in many important points. But in some respects it fails. Electricity, whether positive or negative, can be transferred from one body to another, so that a body may be charged with an excess of electricity of either kind. It is not so with magnetism. Every magnet possesses both polarities to an equal degree, though they may be diffused through portions of its mass of

different extent. A long conductor, exposed to the inductive influence of an electrified body, has opposite electricities developed at its ends. If, now, it be divided in the middle, we obtain the two electricities separate; one half of the conductor possessing an excess of positive, the other of negative electricity. The condition of a magnet in regard to the distribution of its polarities appears to be exactly analogous to that of the conductor; the north polarity seeming to be collected in one half of its length, and the south in the other. We might, therefore, expect that, by breaking the magnet in halves, we should obtain the two polarities separate, one in each portion of the bar. But such is not the case; each half at once becomes a perfect magnet. In Fig. 101, which represents a fractured magnet, the original north pole

Fig. 101.

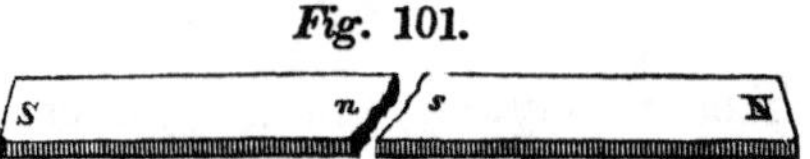

still remains north, but the broken end of the piece has acquired a south pole. The converse occurs with respect to the other portion in which the south pole was situated. These halves may be again broken with the same result; and, in fact, into however small fragments a magnet may be subdivided, each will possess a north and a south pole.

248. Suspend a piece of iron from one pole of a magnet, and bring up to this pole the opposite pole of another magnet. The iron immediately falls; the poles, when in contact, neutralizing each other,

and corresponding to the middle or neutral portion of a magnet. If the piece of iron is nearly as heavy as the pole can sustain, it falls on the mere approach of the other magnet to the pole, and before it touches it.

249. In order to explain the distribution of the magnetic forces and the results of fracture, it is necessary to regard a magnet as made up of minute, indivisible particles of steel, each of which possesses the properties of a separate magnet. Fig. 102 is intended to give an idea of the mode of arrangement of these elementary magnets; they are, of course, so minute as to be beyond the reach of the most powerful microscopes. All their poles lie in the same direction, as indicated by the light and shaded portions in the cut; the light half of each little magnet representing its north pole, and the shaded half its south pole. These poles are distributed through the whole length of the bar; but in consequence of their mutual reactions, the resultant polarities are strongest near the ends of the bar, becoming more and more completely neutralized towards the middle.

Fig. 102.

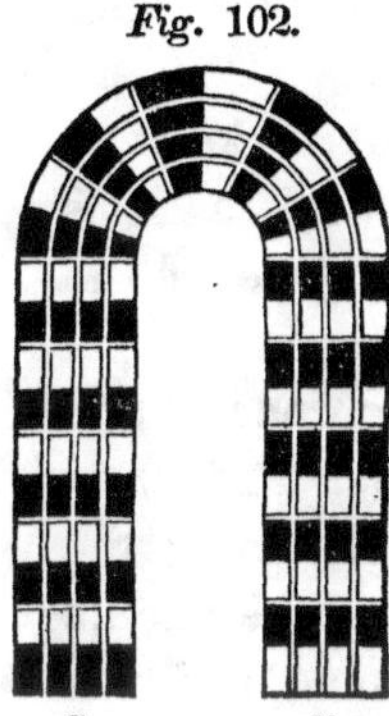

250. Artificial magnets were formerly made by induction from strong magnets previously prepared; the original source of the power being natural magnets, or the magnetism of the earth. When this was the case, it became important to ascertain what arrangements, and what modes of applying a magnet

to a bar or needle, were most efficacious in communicating or developing the magnetic power; and, accordingly, various and complicated arrangements and manipulations for this purpose, are detailed in old treatises on the science. Recently, other and far more powerful means have been discovered for magnetizing bars of iron or steel, by the aid of electro-magnetism; so that the old methods have been, in a great measure, superseded. The induction of magnetism by means of steel magnets is now only employed for needles or small bars.

251. It may, however, be convenient to know a good process for magnetizing (or *touching*, as it is technically called) with steel magnets. One of the simplest and best for a small straight bar is the following: Place the middle of the bar on one of the poles of either a straight or a U-magnet, and draw one end of it over the pole a number of times; the direction of the motion being always from the middle to the end. Then turn the bar in the hand, and pass the other half over the other pole of the magnet in the same way. If the bar is thick, the process may be repeated with its different sides. The end which has been drawn over the south pole of the magnet will now possess north polarity, and the other extremity south polarity. A U-magnet is more readily charged by drawing over it the poles of a steel U-magnet, of corresponding width, in the same manner as in the process, which will be described farther on, for touching with electro-magnets.

252. The magnet which is used to induce mag-

netism loses none of its own power in the process, but often receives a permanent increase by the reaction of the polarities it has induced upon its own. A magnet, whose armature is kept constantly applied, not only retains its power without loss, but acquires a further increase up to a certain limit. That a magnet possesses greater power while exerting its inductive action, may be shown by suspending from one pole of a bar magnet as much iron as it can hold. If, now, a bar of iron be applied to the other pole, the first will be found capable of sustaining a greater weight than before.

II. BY THE INFLUENCE OF A CURRENT OF ELECTRICITY.

253. It has already been stated, under the head of the directive tendency of a magnet in reference to a current of electricity, that a magnetized body, freely suspended within the influence of such a current, tends to assume such a position that the line joining its poles may be at right angles to it. It is also found, that, if any magnetizable body be placed in the vicinity of an electrical current, it acquires magnetism by its influence. This phenomenon is termed *electro-magnetic induction.* The present chapter, and the one referred to above, form together the department of *electro-magnetism.*

254. In Fig. 103 is given a sectional view of a series of four pairs of Grove's battery, whose poles are connected by a copper wire, in which the current

is flowing in the direction of the arrow. A small iron rod, or short piece of iron wire, placed vertically in front of the horizontal portion of the copper wire, becomes instantly a magnet, with its north pole below the wire. The direction of the induced polarity is reversed by transferring the iron rod to the other side of the wire. The rod being placed in a horizontal position above the wire, its north pole is turned towards the observer; if below the wire, it is directed from him. By carrying the rod around the

Fig. 103.

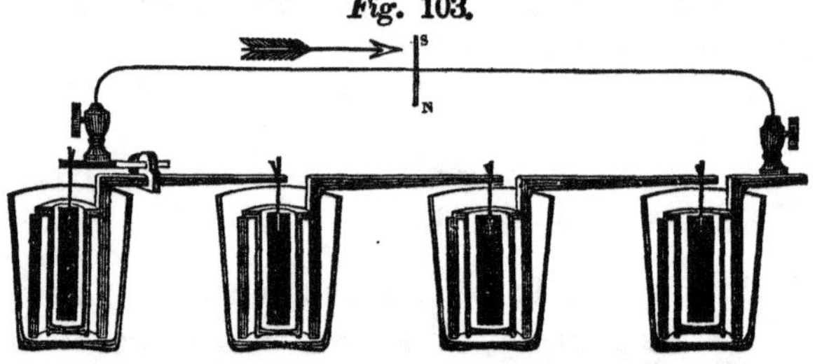

wire, keeping it transverse, it will be seen that the induced poles retain their relative position with regard to the current. As the rod is carried round, the end which was a north pole in the position shown in the cut, remains north in every part of the revolution. The rod attracts iron filings, and acts upon a delicate magnetic needle, brought near it, like a steel magnet; but its polarity instantly disappears, when it is removed from the influence of the current. The induction diminishes as the rod is removed farther from the wire; but it is not interfered with by the interposition of any bodies not susceptible of magnetism.

255. A sewing needle, placed transversely across the wire, becomes magnetic in the same way, and retains a portion of its power when removed. If placed parallel to the wire, it acquires feeble polarity on its opposite sides, instead of in the direction of its length, and probably will not retain it after removal; it being very difficult to maintain this transverse distribution of magnetism in magnets whose length considerably exceeds their diameter.

256. Though the relation between the electric current and the direction of the polarity which it induces is fixed and determinate, yet it is very difficult to express. The same law which has been stated, in § 153, as governing the motion of a magnetic pole brought near the current, regulates the direction of the induced magnetism. It is convenient to reduce this to a rule, by which the position of the poles may be readily found in all cases.

257. The following mode of fixing the rule in the memory, is perhaps the best that has been contrived. First, it is more natural to fix our attention on the current of *positive* than of negative electricity Secondly, in a vertical wire, a *descending* current will occur to us more readily than an ascending one; or, if we imagine ourselves borne along by the current, it would be more natural to conceive ourselves moving with our *feet* foremost; but if, on the contrary, we suppose ourselves to be at rest, we should conceive the current to be passing from our head to our feet. Our *face* would, of course, be turned *towards* the body to be magnetized; we should

attend to the *north* pole in preference to the south; and to our *right* hand rather than to our left. Combining these conditions, then, we may always recollect, that, *if we conceive ourselves lying in the direction of the current, the stream of positive electricity flowing through our head towards our feet, with the bar to be magnetized before us, the north pole of that bar will always be towards our right hand.* If any one of these conditions be reversed, the result is reversed likewise.

258. Though a single magnetic pole is neither attracted nor repelled by a conducting wire (§ 152), the tangential action of the current upon both poles of a needle, situated as in Fig. 103, causes it to approach the wire. This result may be shown by placing the conducting wire in a vertical position, and presenting to it a needle suspended in a horizontal position by a thread. When the needle is removed to some distance, or when very short, the attraction becomes insensible. From the feeble magnetizing power of the wire, this experiment is best performed with a needle previously charged. If the needle is brought up to the wire with its poles in the reverse direction to those of the iron rod in Fig. 104, or if, without changing their direction, it is carried to the other side of the wire, it is repelled by the combined tangential forces.

259. Fig. 104 illustrates the action of the forces in producing attraction and repulsion. W represents a horizontal section of the conducting wire, in which the current is ascending; and N S a magnetic

needle, whose poles are at equal distances from the wire. From W, as a centre, a circle is drawn, passing through the poles. The forces which move the magnet are tangents to this circle, and their directions are indicated by the arrows. It will be seen that the resultant of the forces, acting on each pole, urges the centre of the magnet towards the wire. If the magnet be transferred to the position N′ S′, on the other side of the wire, its centre is urged away from it. The force increases in proportion as the magnet is nearer to the wire.

Fig. 104.

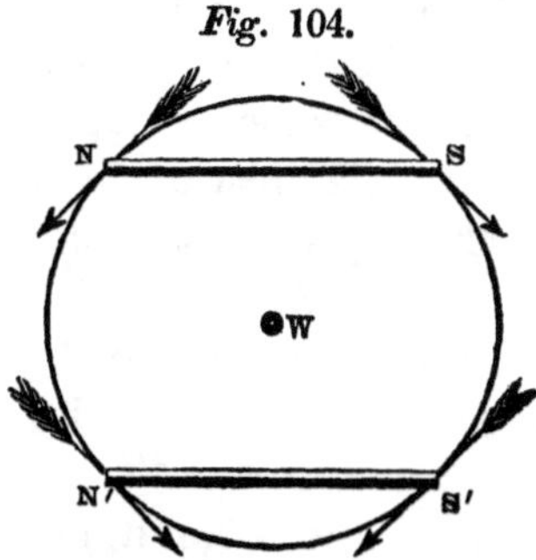

260. When the poles are at different distances from the wire, the resultant of the tangential forces moves the magnet obliquely towards the wire, until its centre comes into contact with it. If there are two currents moving in opposite directions, one on each side of the magnetized bar, and at equal distances from it, their combined action urges the bar forward until its centre comes into the same line with the wires. In Fig. 105, let W be a horizontal section of a wire,

Fig. 105.

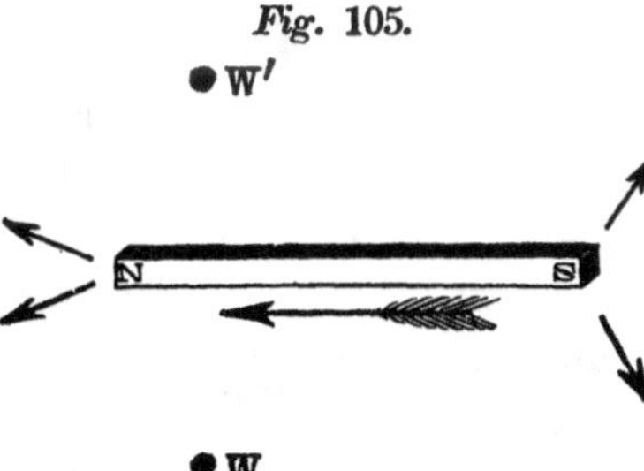

in which the current is ascending, and W′ one in which it is descending. The four small arrows indicate the tangents to the circles, drawn as in Fig 104, from W and W′ as centres. At the south pole of the bar, the direction of the forces is nearly opposite, and they neutralize each other in a great degree: at the north pole, they approach to parallelism, and that pole is urged forward by a force nearly equal to the sum of the two. As this force is opposed only by the feeble one at the south pole, the bar moves in the direction indicated by the arrow below. When the magnet is not equidistant from the wires, it no longer moves in the direction of its length, as in the cut, but its centre is drawn towards the nearest wire, with a force which is accelerated as it approaches.

261. A short copper wire, connecting the poles of a battery, will sustain iron filings, as represented in Fig. 106. The lines of filings have not that bristling, divergent arrangement which they exhibit under the influence of a steel magnet, but adhere equally all around the circumference of the wire; forming circular bands, the particles of which mutually cohere in consequence of each particle becoming a magnet with its poles transverse to the wire. The influence is equal at every part of the length of the wire; hence these

Fig. 106.

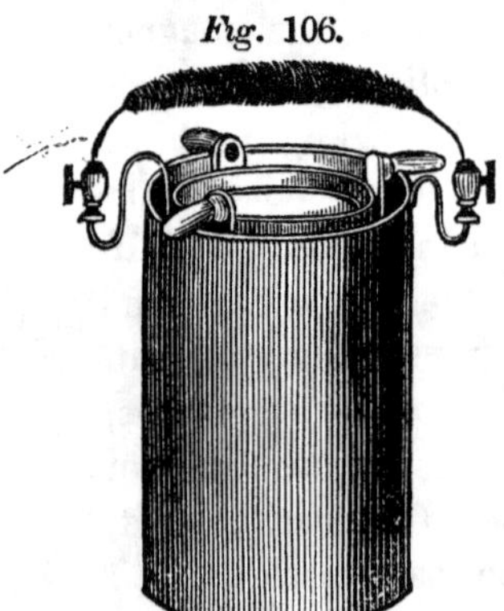

transverse bands, lying in contact with each other, present the appearance of a closely-compacted layer. Whatever form the metal conducting the electricity may have, the filings will always arrange themselves in lines encircling it at right angles to the course of the current. The iron filings fall off when the current ceases to flow; but if steel filings be employed, part of them remain attached, in consequence of the adhesion of the magnetized particles among themselves. A short wire is better for this experiment than one of considerable length; and a battery of some power, such as a series of a few pairs of Grove's, may be used with advantage.

262. Helix, on Stand. — The magnetizing power of the wire is very greatly increased by coiling it in

Fig. 107.

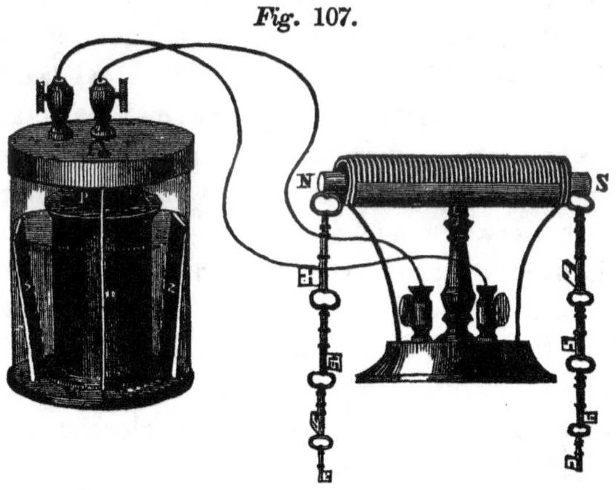

the manner of a corkscrew, so as to form a hollow cylinder, into which the body to be magnetized can be inserted. Such a coil is denominated a *helix*, and is represented in Fig. 107, mounted upon a stand. A single circular turn is more efficient than the straight wire, and each turn adds to the power within a certain limit, whether the whole forms a single layer, or whether each successive turn encloses the previous one in the manner of a spiral. When a helix of great power is required, it is composed of several layers of wire. The wire forming the coil is insulated by being wound with cotton, to prevent any lateral passage of the current.

263. Place a bar of soft iron within the coil, and connect it with the battery by means of the two cups attached to the stand. The two extremities of the bar instantly become strongly magnetic, as will be seen by bringing a key, or other piece of iron, in contact with them. On separating one of the wires communicating with the battery, the magnetic power of the iron bar will be immediately destroyed, and the key will drop. If iron filings, or small nails, are held near one of the extremities of the iron, they are taken up and dropped alternately, as the connection with the battery is made or broken. By bringing a magnetic needle near the two extremities of the bar, in succession, one of them will be found to have north and the other south polarity, and they will attract and repel the poles of the needle accordingly. A bar, temporarily magnetized by an electric current, is called an *electro-magnet.*

264. The following rule indicates the extremity at which the north pole will be found. If the helix be placed before the observer with one of its ends

towards him, and the current of electricity, in passing from the positive to the negative pole of the battery, circulates in the coil in a direction similar to that of the hands of a watch, or the threads of a common screw, then the north pole will be *from* the observer, and the south pole *towards* him. If it passes round in the contrary direction, the poles will be reversed. Or the formula may be stated thus: *The north pole will be at the farthest end of the helix when the current circulates in the direction of the hands of a watch.* By comparing this rule with that given in § 257, for a straight wire, it will be seen that it is directly deducible from it.

265. Place a steel bar, instead of an iron one, within the helix. It acquires polarity somewhat less readily, but the polarity will continue after the connection with the battery is broken. Any small rods or bars of steel, needles, &c., may be made permanent magnets in this way. Bars of iron, or steel, brought near the outside of the helix, obtain polarity in a much feebler degree. An iron tube does not become perceptibly magnetic when a current is passed through a helix placed within it, though it becomes strongly so when enclosed in a helix sufficiently large to admit it. If two soft iron bars are inserted in the helix, at the opposite ends, in such a manner as to have their extremities in contact in the middle of the helix, they adhere with more force than when one is within and the other not.

266. Place the helix with its axis vertical, and a small rod of iron or steel within it. If it be now

connected with the battery, it may be raised from the table without the bar falling out; the tendency of the helix to keep the bar within it overpowering its gravitation. A small steel bar, merely allowed to fall through the helix, while the current is flowing, acquires a considerable degree of magnetism.

267. If a needle, or a small bar of steel, previously magnetized, is placed in the helix, in such a position as to bring its north pole within the south pole of the helix, as indicated by the rule in § 264, its poles will be reversed, the end which was previously north becoming a south pole. When of very hard steel, and too large in proportion to the power of the coil, its magnetism may merely be diminished.

268. A small magnetic needle, suspended by a thread near the helix, will be drawn within it, its north pole entering the south end of the helix, or its south pole the north end. If the magnet be placed within the helix, in a contrary direction, its north pole entering the north end, it will be repelled, and then, revolving without the helix, will return and enter by the other pole. This effect will take place, unless the electro-magnetic power of the coil is sufficient to reverse the poles. When the needle has entered with its poles corresponding in direction with those of the helix, the action of the coil tends to keep it in the middle of its length, though not in the line of its axis.

269. A magnetized needle, placed exactly in the axis of a vertical helix, would retain this situation, if undisturbed, being acted upon equally by the

forces around it. But it is a position of unstable equilibrium, and if displaced from it in the slightest degree, the needle is drawn towards that side of the coil which happens to be nearest.

270. Let a short helix, such as is described in § 271, be laid on a table, with its axis vertical, and a short and strongly magnetized needle of very hard steel be introduced into it with its poles in the reverse direction to those of the coil. If its polarity is not reversed by the power of the current, it will be found to stand erect in the axis of the helix, being repelled equally from every side, instead of being attracted to every side, as is stated in § 269 to occur when its poles correspond in direction with those of the coil.

271. Heliacal Ring. — This is a short helix, consisting of several layers of wire, and with a large central opening. It is represented at C, in Fig. 108. The ends of the wire are left free, in order to be inserted into the cups of the battery. At D are shown two semicircular pieces of soft iron, provided with handles for pulling. The handles are attached to the semicircles, by ball and socket joints, to prevent them from being twisted or wrenched by irregular pulling. When connection is made with the battery, and the semicircles are passed within the coil, as shown in the cut, they adhere with very considerable force. The induction of magnetism in these

Fig. 108.

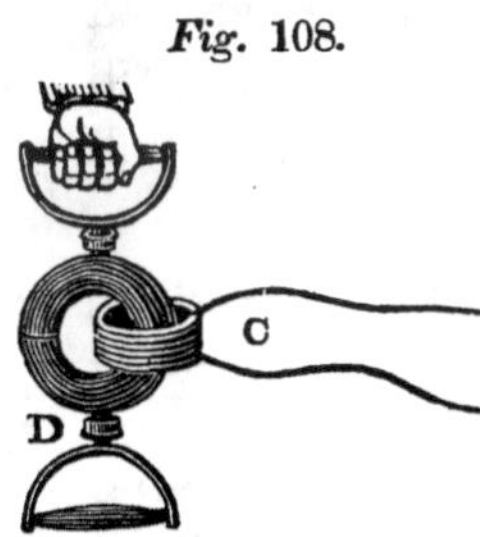

semicircles, by means of the current from a thermo-electric battery, has been mentioned in § 129.

272. With thicker semicircles, such as are represented in Fig. 109, the induced magnetism is sufficient to support a hundred weight or more, even with a small battery. The mutual attraction between the semicircles when near each other, but not in actual contact, is comparatively very feeble. If the flow of the current in the coil is stopped while they are applied to each other, they still continue firmly attached; but if once separated, will not adhere again.

Fig. 109.

273. A heliacal ring, with a smaller central opening, will sustain a large mass of tacks or brads while the current is flowing through it, as shown in Fig. 110. They adhere together in rings, surrounding each portion of the circumference, as the filings enclose the single wire, in Fig. 106. The poles of a magnetic needle brought near the coil are attracted and repelled as if its axis were a magnet. An iron bar becomes strongly magnetic when passed half way through it. In § 185 is described the polarity exhibited by a coil when it has freedom of

motion. The polarity of the helix bears so close a resemblance to that of a magnet, that it is usual to

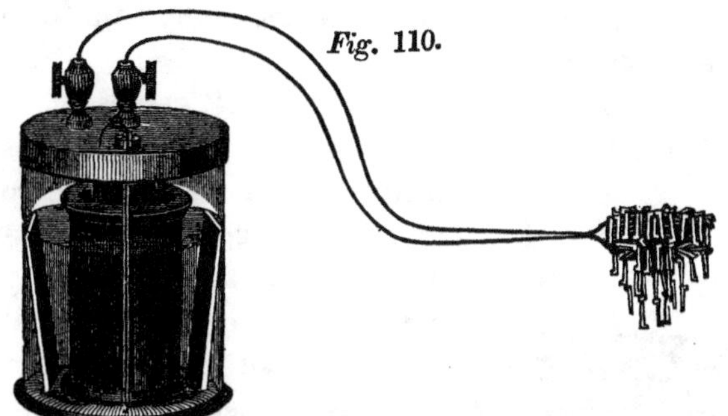
Fig. 110.

speak of it as magnetic, and to give the names of north and south poles to the extremities of its axis.

274. In Fig. 111 are represented, at *a a*, two double cylinders of iron, enclosing a coil. A sectional view of one of them is given separately at A. It contains a cavity in the form of a hollow cylinder, adapted to receive one half of the coil seen at C, the remaining half passing into the other cylinder when they are fitted together. A longitudinal opening on one side of the cylinders allows the wires of the coil to pass out. In this arrangement, the cylinders adhere with great force when in contact; but as soon as the current ceases, their magnetism instantly dis-

Fig. 111.
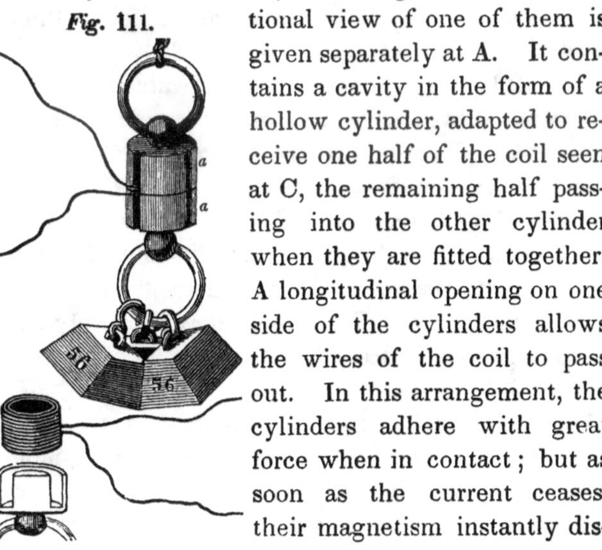

appears, and the adhesion does not continue as with the semicircles. When one of the cylinders is passed over the coil, the part exterior to the coil exhibits no sensible polarity.

275. Ribbon Spiral. — Fig. 112 represents a strip of sheet copper, coiled into a spiral. This instrument is described here in consequence of its possessing considerable magnetizing power, though its principal uses will not be mentioned till the subject of electro-dynamic induction comes under consideration, in book III, chapter I. The copper ribbon may be an inch wide and one hundred feet long, the strips being cut from a sheet, and soldered together. Being then wound with strips of thin cotton cloth, it is coiled upon itself, like the mainspring of a watch; instead of covering it with cotton, it may be coiled with a strip either of cotton or list intervening. Two screw-cups are soldered to the ends of the ribbon; the internal end, for convenience, is brought from the centre, underneath the spiral, to its outside, care being taken to insure insulation where it passes the edges of the ribbon. The whole may be firmly cemented together, if desired, by a solution of shellac in alcohol.

Fig. 112.

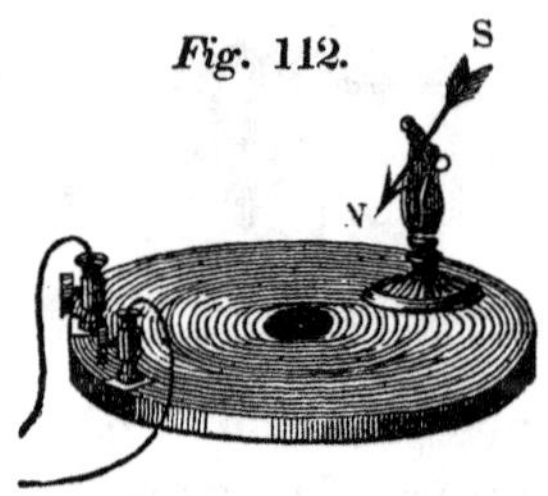

276. The spiral being connected with the battery, its two faces exhibit strong polarity: a dipping needle, placed on any part of its surface, or near it, will always direct one of its poles towards the centre,

as seen in the cut, where a dipping needle, N S, is represented on the spiral. On reversing the battery current, the other pole of the needle will turn towards the centre. If the spiral is fixed in a vertical position, a horizontal magnetic needle may be used with the same result. When brought near to one side of the coil, it will be found to direct its north pole constantly towards the centre; when on the other side, its south pole. When either the horizontal or dipping needle is placed near the outside, with its centre of motion in the same plane as the spiral, neither pole will be directed towards the centre, but the magnet will place itself at right angles to the plane of the spiral.

277. The magnetizing power of the spiral may be shown by connecting it with the battery, and placing a rod of iron or steel in the central opening, or upon it in the direction of a radius, when the iron becomes temporarily magnetic, and the steel permanently so. If the bar, when laid upon the coil, extends across the central opening, both ends will become similar poles, and the part over the centre, a pole of the opposite denomination.

278. Magnetizing Helix. — For the purpose of experimenting on the magnetic power developed by the electric current in small bars of iron or steel, a helix of the form represented in Fig. 113, is well adapted. It consists of a number of layers of wire, and has a small central opening. An iron bar passed within it, becomes strongly magnetic. When the coil is in a vertical position, the iron bar is sustained

within it in consequence of the force with which it is drawn towards the middle of the coil. With a large battery, a considerable weight may be suspended from the bar, as represented in the cut, and the whole will be sustained without any visible support. The helix represented in Fig. 107, will lift a smaller rod in the same manner.

Fig. 113.

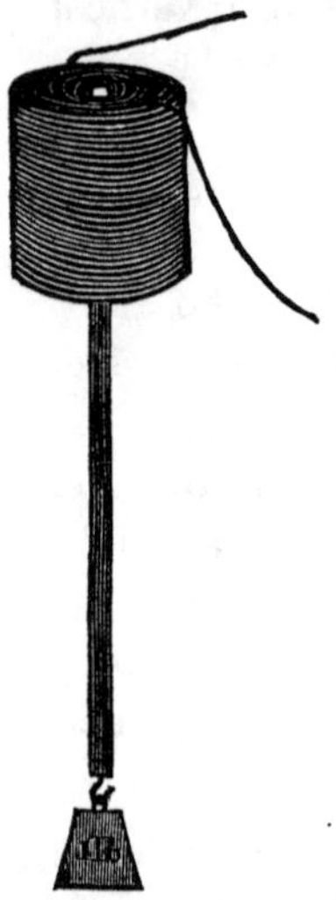

279. The action of the coil is the same, except in amount, as that of a single circular turn of the wire. At any two points of the circle diametrically opposite, the directions of the current are also opposite. Such points may be represented by the sections of the wires, W and W′, in Fig. 105, the action being precisely the same as explained with reference to them. The resultant of the forces exerted by all the points, tends to bring the centre of the magnetized bar within the circle. The action of all the circles of which the helix is composed, draws the bar into it until its middle lies within the middle of the helix, in which position only can the forces neutralize each other. The terms *axial force* and *axial motion* are used to designate this peculiar force and the motion occasioned by it.

280. Axial Bell Engine.—Fig. 114 represents an instrument in which the axial force of a vertical helix raises an iron rod whose centre is below

that of the coil. This motion is communicated to a hammer, which strikes a bell. A wire, fixed to the iron rod below, rests on a spring near B. When the rod is lifted by the current in the coil, C, its upper end raises the handle of the hammer. A guiding wire, R, keeps the rod vertical. The circuit is broken at B, by the lifting of the handle, and is renewed when the rod falls.

Fig. 114.

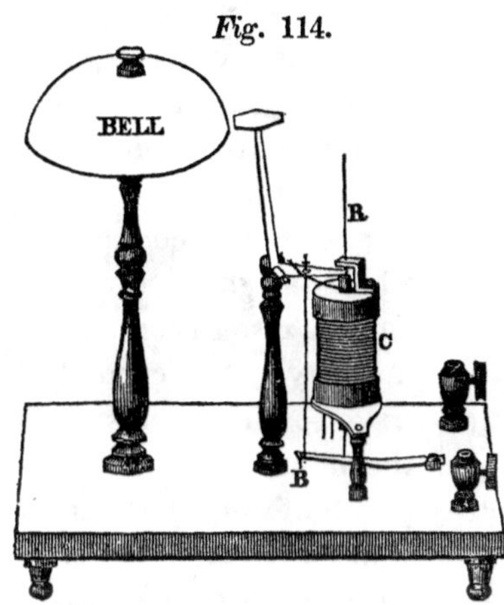

281. Axial Revolving Bar. — In the instrument represented in Fig. 115, an iron bar is suspended without visible support by the action of a current flowing in the helix which surrounds its upper half, the bearings above and below serving only as guides. The bar is made to rotate by transmitting the battery current through either half. This mode of suspending the rotating bar was first suggested by Dr. Boynton. The screw-cup A, on the stand, connects with the brass pillar, and thence with one end of the coil. The other end of the coil dips into mercury, contained in a circular cistern of ivory. This cistern is supported below the helix by an arm attached to the pillar, and has an opening in its centre to allow the bar to pass through. A bent wire, pro-

jecting from the middle of the bar, conveys the current from the mercury through its upper half to the cup B. Some non-conductor of electricity, interposed at I, insulates B from the arm which supports it. If A and B are connected with the battery, and the cistern contains mercury, the current traverses in succession the helix and the upper half of the bar. Since the bar becomes an electro-magnet by the influence of the coil, it rotates rapidly on its axis, in the same manner as the magnet described in § 167. The cup C connects with the lower end of the bar, and, by uniting A and C with the battery, the current traverses the lower half, also producing rotation.

Fig. 115.

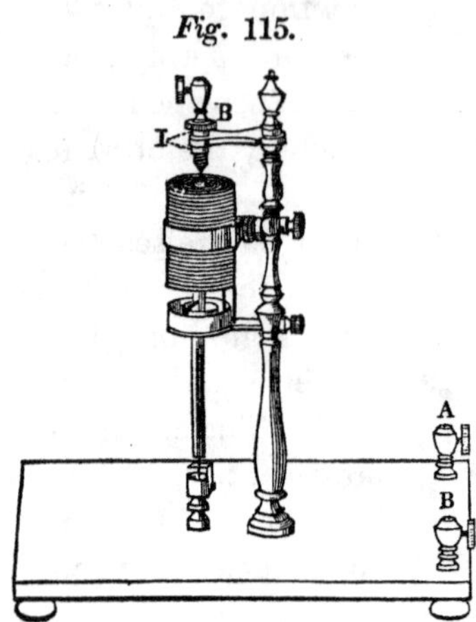

282. In Silliman's Journal for March, 1847, will be found a figure and description of a similar instrument, contrived by Dr. Page, in which no mercurial connections are used. The bar is sustained without visible support, as in Fig. 115, and the current is conveyed by a wire tipped with silver, which plays upon a silver ferule on the middle of the bar.

283. Inclined Revolving Bar.—This instrument, represented in Fig. 116, is designed to show

the change produced in the direction of rotation by altering the course of the current. The helix, which surrounds the lower end of the bar, is represented in section, in order to show a wide glass tube within, which can be filled with mercury. The iron bar rests in an agate cup at the bottom of the tube. Its upper end connects with a brass cup on an ivory arm proceeding from the upright brass pillar. This cup may be united by a bent wire with a similar one on the pillar, as represented in the cut. When the screw-cups on the base board are connected with a battery, the current passes from one of the cups up the pillar and along the bent wire at the top; thence it passes down the inclined bar. A little mercury in the glass tube conveys the current to the brass at the bottom of the tube. It then passes through the helix to the other screw-cup. The bar now rotates on its axis; a circular card attached to it renders the motion perceptible at some distance.

Fig. 116.

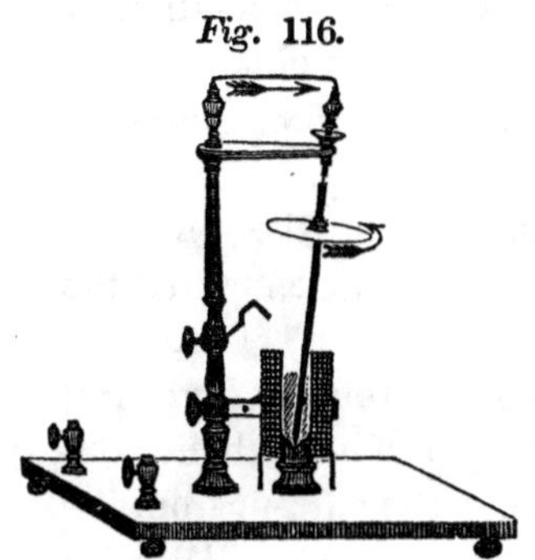

284. The course of the current is changed by re moving the wire at the top of the pillar, and, having filled the glass tube with mercury, dipping into it another bent wire, fixed by a binding-screw to the middle of the pillar. The inclined position of the bar is adopted to prevent this wire from accidentally

obstructing its movement. Only the lower part of the bar is now traversed by the current, which divides itself between that and the mercury. A rotation of the bar in the opposite direction is thus obtained.

285. Inclined Bar revolving round Conductor. — The instrument represented in Fig. 117 is similar to the one last described, except that the inclined bar of iron revolves round a vertical copper wire, which conveys the current from the mercury to the pillar, with which it is connected by a brass arm. The iron bar rotates on its axis in the opposite direction to that of its revolution around the wire. A revolution of the mercury may also be observed in the same direction as that of the iron around the wire. The glass tube requires to be nearly full of mercury to produce contact with the vertical conducting wire, which enters a short distance only within it. An ivory ring, surrounding the wire, insulates the little arm by which the bar is sustained in its inclined position, so that the current does not traverse the upper part of the bar. When enough of the mercury is removed to interrupt contact between itself and the vertical wire, and the bent wire is inserted so as to convey the current from the middle of the pillar, the iron rod changes the direction of its motion.

Fig. 117.

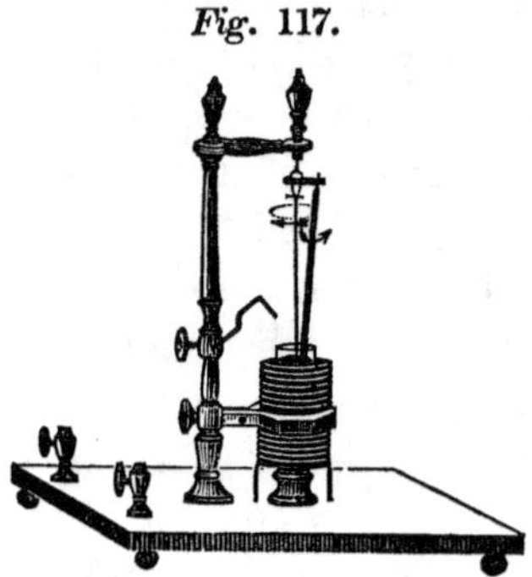

286. A short iron rod, placed in the glass tube instead of the inclined bar, floats in the mercury. When the current is passed through the vertical wire, the rod is drawn down into the tube, and revolves around the wire and on its axis. By covering the rod with sealing-wax, it no longer conveys any part of the current, but is carried round by the rotation of the mercury. If the vertical wire is so arranged as to have freedom of motion, it revolves in the same direction as the iron bar, the two keeping opposite, and moving round a common centre.

287. Axial Revolving Circle. — By employing two curved helices, the axial motion of an iron bar may be converted into a direct rotation. In Fig. 118 is seen a revolving circle, composed of two semicircular bars, the shaded one of iron, the other of brass. This circle is supported by four grooved friction rollers, so as to move freely within the two coils. Its circumference is cut into teeth, and its motion is communicated to two smaller toothed wheels, on the axis of one of which is the break-piece. When the current passes, one of the helices becomes charged, and draws the iron bar within it. As soon as its cent e reaches that of the coil, the

Fig. 118.

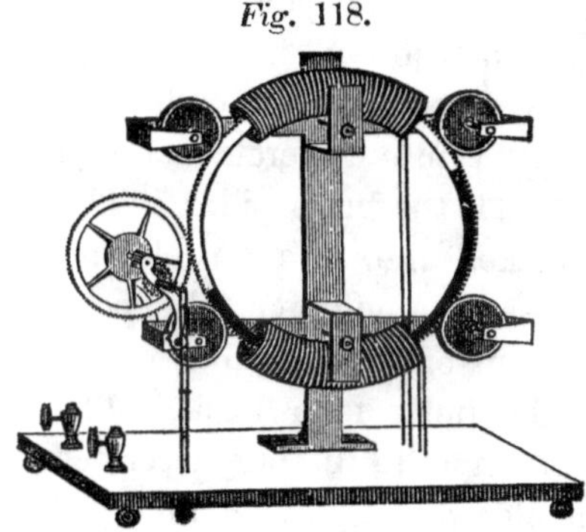

current is interrupted; and, as it moves on by its acquired momentum, the action of the break-piece causes the other helix to be charged, and the iron to be drawn into it.

288. **Vibrating Helix.**—Instead of a bar moving within the helix, the latter may be made to move over the bar. In the instrument represented in Fig. 119, there is a curved bar of iron of the U form, and the helix is so suspended as to pass along one of its poles. When the coil hangs freely, one of its ends dips into a mercury cup on the brass pillar. The action of the current makes the iron an electro-magnet, and the helix passes along over the pole. This motion lifts the wire out of the mercury, which breaks the circuit, and the helix falls back. The instrument is similar in principle to the one described in § 187, except that an electro-magnet is employed instead of a steel magnet.

Fig. 119.

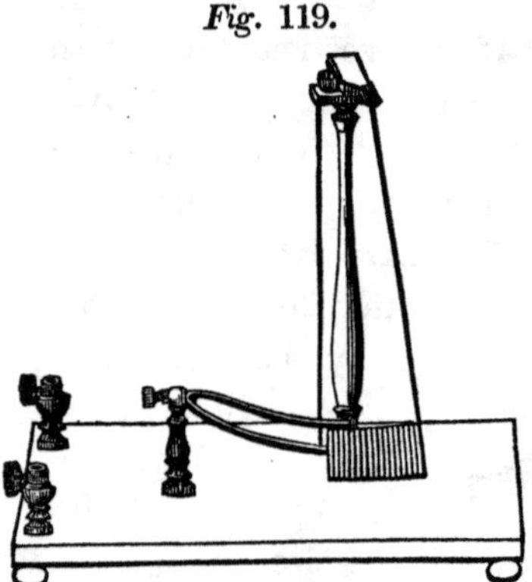

289. In place of a straight iron bar moving within a helix, a bar having the form of the letter U may be employed with two coils placed side by side. In Fig. 120, C C are the two helices, secured together by a brass clamp, not represented; and M is the bent iron bar. Suppose the bar to be resting on a table

in the position indicated by the dotted lines in the cut, with its poles within the double helix, which is

Fig. 120.

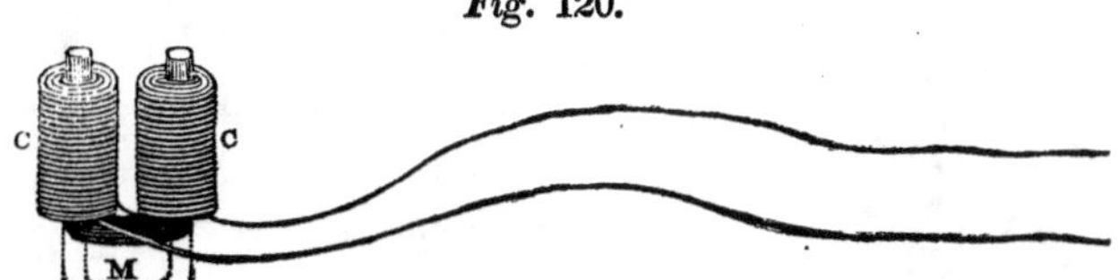

raised a little from the table. When the battery connections are made, the bar instantly rises until its bend reaches the helices. The movement is more powerful than that of a straight bar.

290. Double Axial Bell Engine. — In the instrument represented in Fig. 121, motion is produced in the manner just described. The force with which the bar is drawn up is increased by its attraction for a straight armature fixed above the coils. The bent bar, becoming an electro-magnet, moves towards the armature, as the latter would do, if free to move, towards a fixed magnet. Near B is a break-piece, so arranged that the current is interrupted when the bend of the bar reaches the helices, causing it to fall back. As the electro-magnet rises, it communicates motion to a hammer which strikes the bell.

Fig. 121.

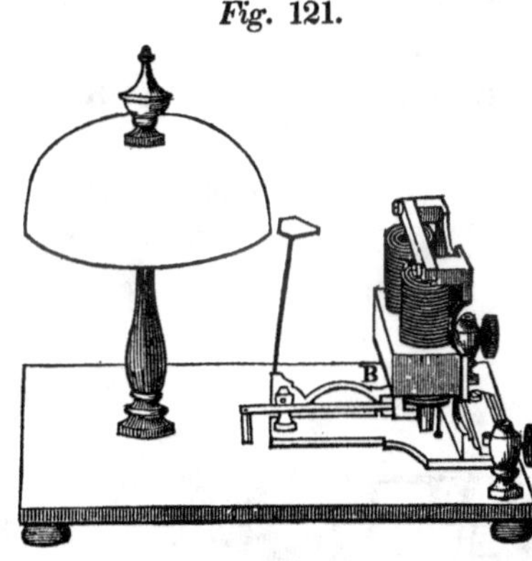

291. Upright Axial Engine.—In the instrument represented in Fig. 122, the alternating movement of two bent bars is conveyed by cranks to a balance-wheel above the helices. As the wheel revolves, the current is conveyed to each double helix in succession, by means of a break-piece on its axis. This is similar in construction to that described in § 187, except that the two wires which press on the break-piece are connected respectively with one end of each double coil, while their other ends both connect with the axis.

Fig. 122.

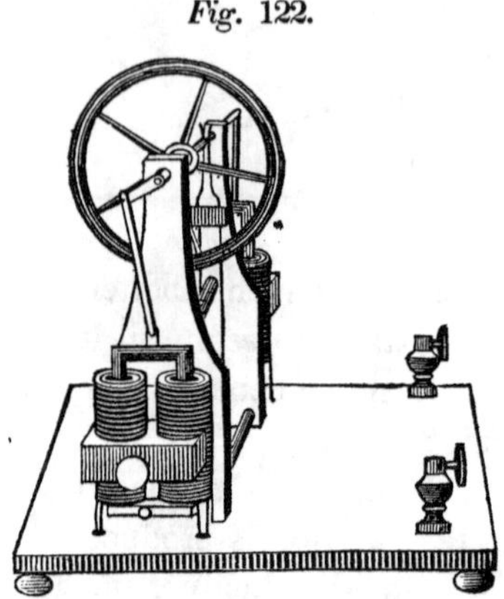

292. Horizontal Axial Engine.—In the instrument represented in Fig. 123, the two double helices

Fig. 123.

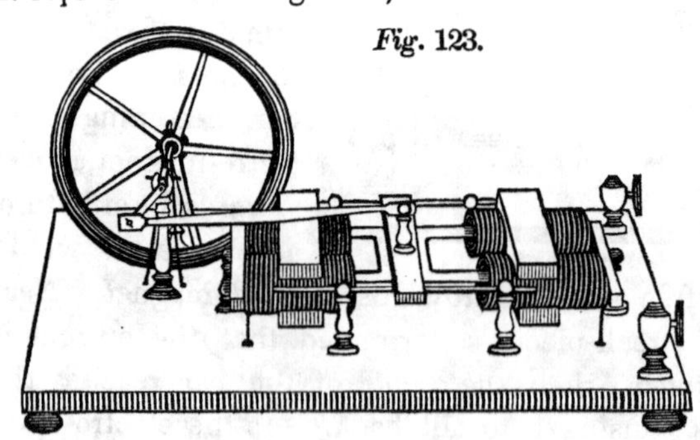

are fixed horizontally in the same line. The iron bars

are connected at their bends by a transverse bar, through the ends of which pass guiding rods. These slide in the tops of the small pillars, and keep the bent bars in place. On the shaft of the balance-wheel is the break-piece.

293. Double Beam Axial Engine. — An instrument in which there are two horizontal beams, each

Fig. 124.

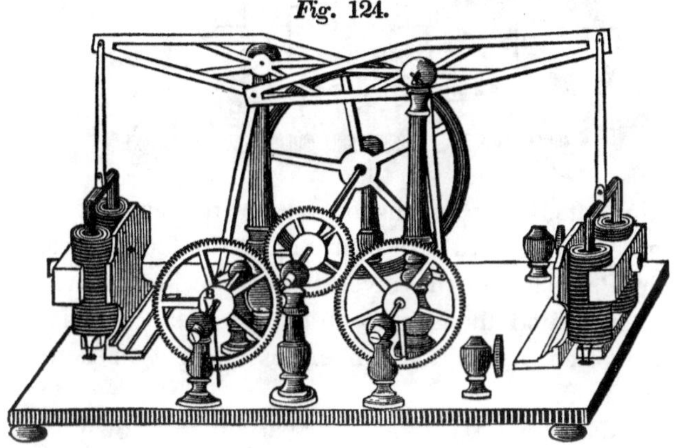

connected with a bent bar moving in an upright double helix, is represented in Fig. 124. The motion of the beams is communicated to a single fly-wheel, by means of toothed wheels moved by cranks.

294. Single Beam Axial Engine. — Fig. 125 represents an instrument in which a reciprocating motion is communicated to a horizontal beam, by the alternate movement of two bent bars. The iron bars are not attached to the ends of the beam, but simply rest upon them. When the current passes, one of the bars is lifted, and the weight of

the other depresses that extremity of the beam on which it rests. There is a break-piece of peculiar con-

Fig. 125.

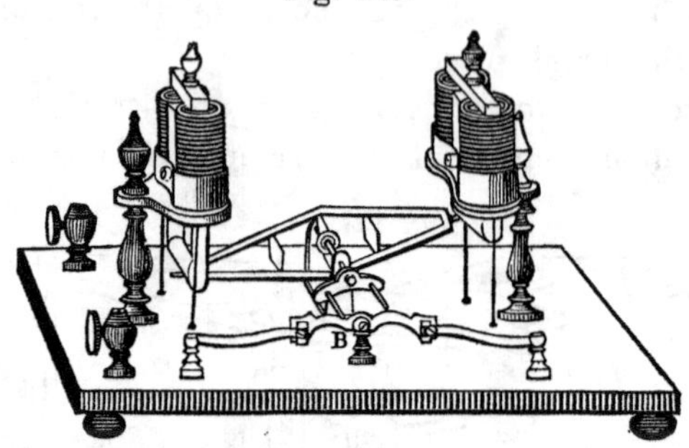

struction at B, by means of which the current is alternately transmitted to each of the double helices. To Dr. Page is due the invention of the first instruments in which the source of motion was the axial force exerted by helices on either straight or bent bars of iron.

295. Electro-Magnet. — A bar of iron, wound with insulated wire, so as to be enclosed in a helix, is termed an *Electro-Magnet* During the passage of an electric current along the wire, the bar exhibits a remarkable degree of magnetic power, far superior to that of a steel magnet of the same size. In all the instruments in which a current is transmitted through a coil surrounding an iron bar, the iron becomes temporarily an electro-magnet.

Fig. 126.

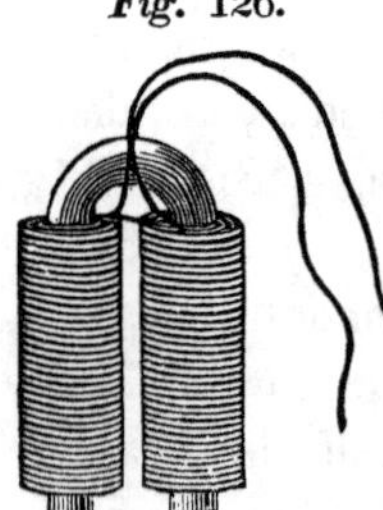

For the purpose of showing its lifting power, the electro-magnet is made of the U form, (Fig. 126.) The battery with which it is used should be in proportion to the length and fineness of the wire, in order to obtain the greatest power. With a thick and rather short wire, a battery of one or two pairs will be preferable. With a long and fine wire, a number of pairs should be employed. A small electro-magnet will sustain a large mass of iron nails or filings about its poles, which fall when the flow of the current is stopped.

296. ELECTRO-MAGNET, IN FRAME. — Fig. 127

Fig. 127.

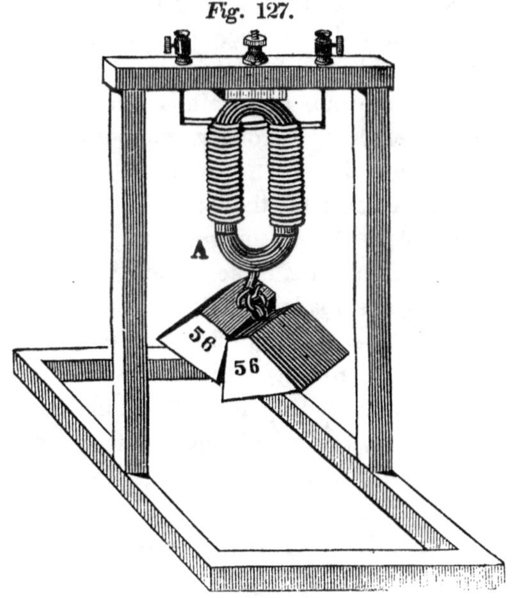

represents an electro-magnet, fixed in a frame, for the

purpose of supporting heavy weights. The armature, A, consists of a semicircular piece of iron. With the electro-magnet, this form is preferable to a straight bar. If the iron of the magnet is soft and pure, its magnetic power is immediately communicated and lost, according as the connection with the battery is made or broken. When, however, the armature is applied to the poles, and the flow of the current stopped while it is attached, it will continue to adhere for weeks or months with great force, so as to be able to sustain one third or one half as much weight as while the current was circulating. But if the keeper be once removed, nearly the whole magnetism disappears, and the magnet, if of good iron, will not even be able to lift an ounce. The purest iron retains, for a time, a certain amount of magnetic power, after being strongly magnetized, even when an armature is not applied. This amount increases with the size of the bar.

297. The poles of the magnet are, of course, reversed by changing the direction of the current. When the change is made rapidly, while the armature is applied, it does not fall off. If a considerable weight is suspended from it, it may fall a very slight distance, and be attracted again. The time required to destroy the previously existing polarity, and to renew it in the opposite direction, is exceedingly short.

298. Electro-Magnet, in Case. — An electro-magnet of considerable power is represented in Fig. 128, fixed horizontally in a case. The poles project

from one end, and a semicircular armature may be applied to them, as in the cut. At the other end of

Fig. 128.

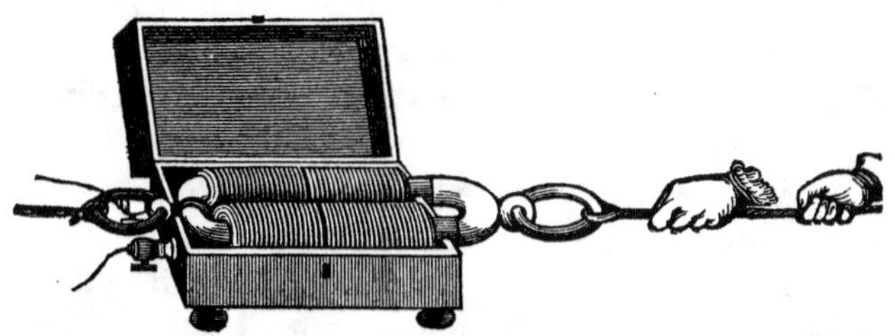

the case are seen two screw-cups, for making connection with the battery. There are two strong rings, one attached to the bend of the magnet, and the other to the armature. The magnet, with its case, may be suspended by the former ring and weights hung from the latter.

299. Very large electro-magnets have been made to lift 3000 lbs. The proportion of the power to the weight of the bar increases as their size is diminished, up to a certain limit; and a small electro-magnet may be made to sustain 400 or 500 times its own weight, excluding that of the coil. Increasing the battery current does not increase the magnetism indefinitely. Small bars acquire their full power, oı become saturated with magnetism, with a moderate battery. A current from the thermo-electric battery (Fig. 41), when transmitted through the wires of an electro-magnet, induces a considerable charge of magnetism.

300. An electro-magnet, like the steel magnet

exerts its attractive force through intervening substances; and the phenomena are more striking with the former, in consequence of its greater power. It will often be able to lift its armature, with a plate of glass interposed; and when a few thicknesses of paper only intervene, a considerable additional weight will be supported.

301. Electro-Magnet, with Three Poles. — This instrument, represented in Fig. 129, consists of an iron rod, wound with insulated wire, which is carried in one direction around half the length of the rod, and then turns and is wound in the other direction. The effect of this arrangement is, that, when connection is made with the battery by means of the screw-cups on the stand, the two extremities of the bar become similar poles, while the middle acquires a polarity opposite to that of the ends. The middle, as well as the ends, will sustain a considerable weight of iron. By reversing the direction of the current, all the poles are reversed. The arrangement of the poles may be shown by passing a magnetic needle along the bar.

Fig. 129.

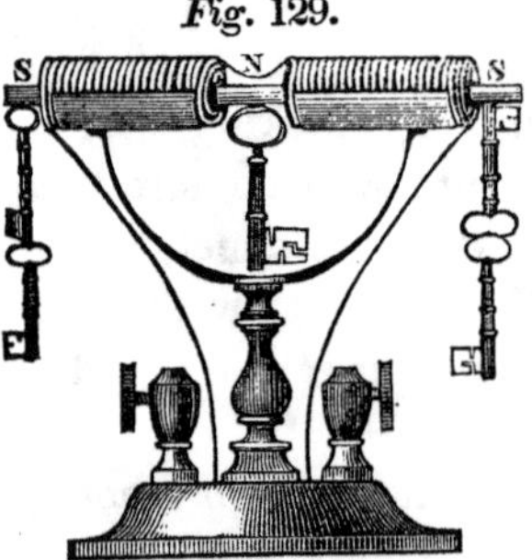

302. Communication of Magnetism to Steel by the Electro-Magnet. — The great power possessed by the electro-magnet renders it peculiarly fitted for inducing magnetism in steel hence it is very con-

venient for charging permanent magnets. A short steel bar, if applied like an armature to the poles of an electro-magnet of the U form, will become strongly magnetic, the end which was in contact with the north pole acquiring south polarity. A longer bar may be charged, by employing the same process that has been described in § 251, for *touching* by steel magnets.

303. Bars of the U form are most readily magnetized by drawing them from the bend to the extremities across the poles of the U-electro-magnet, in such a way that both halves of the bar may pass at the same time over the poles to which they are applied. This should be repeated several times, recollecting always to draw the bar in the same direction. Then, if it has a considerable thickness, turn it in the hand, and repeat the process with its opposite surface, keeping each half applied to the same pole, as before. In Fig. 130, the arrow indi-

Fig. 130.

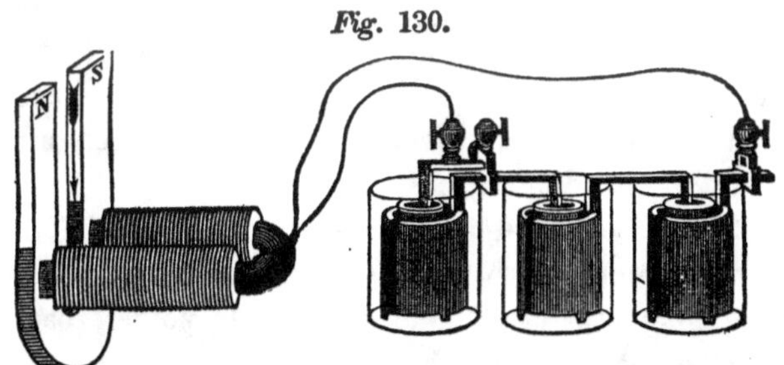

cates the direction in which the motion should take place. Of course, the result will be the same, if the steel bar is kept stationary, and the poles of the electro-magnet are passed over it in the reverse direction.

304. In order to remove the magnetism of a steel magnet of the U form, it is only necessary to reverse the process; that is, to place one of its poles on each pole of the electro-magnet, and draw it over them in the opposite direction to that of the arrow in Fig. 130. In this way, the magnet may be so completely discharged, as to be unable to lift more than a few iron filings.

305. A bar magnet may be deprived of its magnetism, in a great degree, by passing the north pole of an electro-magnet over it, from its south pole to its middle, and then lifting it off perpendicularly; if, then, the south pole be passed in the same manner over the other extremity of the steel bar, it will be found to have lost the greater part of its polarity. If necessary, this process may be repeated several times. A still more effectual mode is to make use of two electro-magnets; place the north pole of one on one end of the bar, and the south pole of the other on its other extremity, and draw the poles along the bar till they meet at its middle; then lift them off perpendicularly.

306. The electro-magnet described in § 298 is very convenient both for communicating and removing magnetism. On account of its power, bars of considerable size may be fully charged by it. If the steel magnet has not the same, or nearly the same, width, from pole to pole, as the electro-magnet, it can be charged by drawing each half separately

over the proper pole of the electro-magnet, in the manner described in § 251, for bar magnets.

307. Magnetometer. — A simple application of the electro-magnet brings us at once to a very useful instrument, the *Magnetometer*, represented in Fig. 131. Its use is to measure the magnetizing power of galvanic currents. The form in which it is now described is new. It consists of a vertical electro-magnet, of the U form, with an armature above it, attached to the short arm of a balanced lever. The long arm of the lever is graduated decimally to measure, by means of weights of from 100 grs. to 10,000 grs., the force required to detach the armature from

Fig. 131.

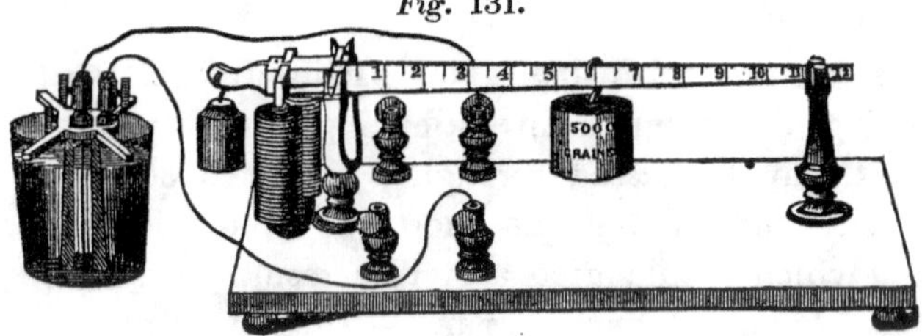

the electro-magnet when connected with the battery whose power is to be determined. The lever is supported on an axis with knife-edge bearings. The armature may also be suspended on knife edges attached to the beam. On the under surface of the armature is brazed a thin plate of brass, to prevent its adhesion to the poles. A difference of magnetizing power of 10 grs. can be estimated in a series extending from 100 grs. to more than 100,000 grs., or the limit of saturation of the magnet. Two sets of screw-cups will be seen on the board; one of these is connected with a short coil round the magnet, the other with a long coil. By making this long coil of fine wire, the instrument compares currents differing in their intensity. Two batteries are first estimated, as to quantity, by their magnetizing power through the short coil. Their relative intensity is then shown compared with their quantity, by their magnetizing power through the long coil, their intensity being in mathematical proportion to the conducting power of the wire for each, and therefore to the amount of electricity which passes. In comparing the power of different batteries, — a matter now of some practical importance, — this instrument gives a rapid and uniform result.

308. Instead of using a long coil of wire, surrounding the magnet, in estimating intensity, the current may be passed through a detached coil of fine wire, and through the short coil of the instrument, which would give a similar result.

309. Axial Magnetometer. — Another form of

Fig. 132.

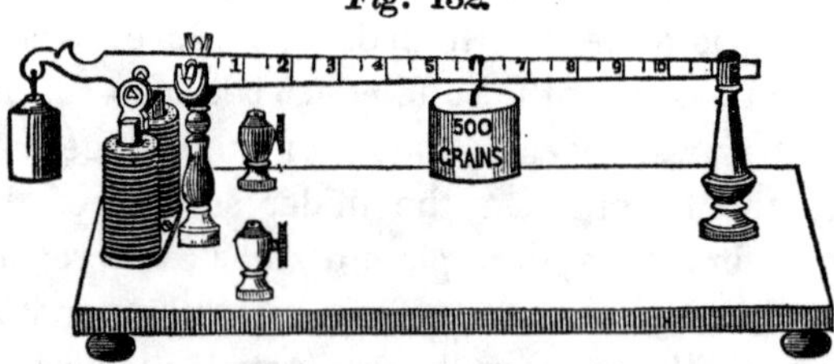

the magnetometer is represented in Fig. 132. In this, the axial attraction of a double helix is made use of, as in the Axial Galvanometer invented by Dr. Page, and described in Silliman's Journal, vol. xlix. p. 136. In other respects, the construction is entirely different from his. A double coil of fine wire may be added, as in Fig. 131, to compare intensity currents.

310. Electro-magnetic Telegraph. — Within a few years, the electric telegraph, the most important application of galvanism ever made, has been contrived and brought into practical use. A slight sketch of the history of this invention, drawn partly from Vail's book, "The American Electro-Magnetic Telegraph," and partly from the original sources, will be of interest here as an introduction. This is the more needed, as there has been great confusion as to the originators and inventors of the telegraph. We find that the first electrical telegraphs were put in operation by Lomond, in 1787; by Reizen, in 1794; and by Salva and Belancourt, in 1798. In these, the common electrical machine was used, with which, of course, no important result could be obtained. In 1809, Sœmmering laid the foundation of the galvanic telegraph, by proposing to employ, with the Voltaic pile, a wire for every letter in the alphabet, terminating in gold pins inserted in a glass cell containing water, the letters being indicated by the evolution of oxygen and hydrogen from the pins. A similar plan was proposed by Dr. Coxe, of Philadelphia, in 1816. In the same year, the use of machine electricity for the telegraph, was revived by Ronalds, in England.

311. On the discovery of electro-magnetism, in 1819, Ampere very early proposed the deflection of the needle in place of the decomposition of water for indicating the passage of the current. In the treatise on new discoveries in electricity, by Ampere and Babinet, published in Paris, in 1822, the telegraph of Wheatstone is anticipated by fifteen years, in all essential details, with the exception of the greater number of wires required in the former. Here it is necessary to draw a line between the proposers and introducers of an invention. The claim of originality belongs to the former, but public gratitude is usually accorded to the latter. The deflection of the needle for telegraphic purposes was subsequently referred to by many other writers; but it seems first to have been introduced into practice by Schilling, in Russia, at the end of 1832; by Gauss and Weber, at Gottingen, in 1833; in Vienna, in 1836; and, finally, on a large scale, by Wheatstone, in England, and by Steinheil, at Munich, in 1837, or soon after. The credit of the construction of the galvanic telegraph belongs thus to Schilling, Steinheil, and Wheatstone, by the latter of whom, with some of his English coadjutors, many of the practical difficulties in the modes of transmitting the current were overcome. The invention of Grove's battery was also an important era in the introduction of the telegraph.

312. We come now to the electro-magnetic telegraph, which, in its beautiful simplicity and efficiency, surpasses all others yet introduced into practice

Barlow, in England, seems to have made some early suggestion of this kind; but it was not until 1830, on the construction of the first powerful electro-magnets, by Professor Henry, of Princeton, N. J., that this form of telegraph became possible; and in his first paper on the results of these experiments, in some of which long wires were used, he at once applies the new facts to the telegraph. The present form of the American telegraph is claimed to have been suggested in 1832, by Dr. C. T. Jackson, and by Professor Morse. It was finally matured by Professor Morse, and introduced by him between Baltimore and Washington, in 1844, after a delay before Congress of more than six years.

313. ELECTRO-MAGNETIC TELEGRAPH. — Fig. 133 represents the recording part of the telegraph. It

Fig. 133.

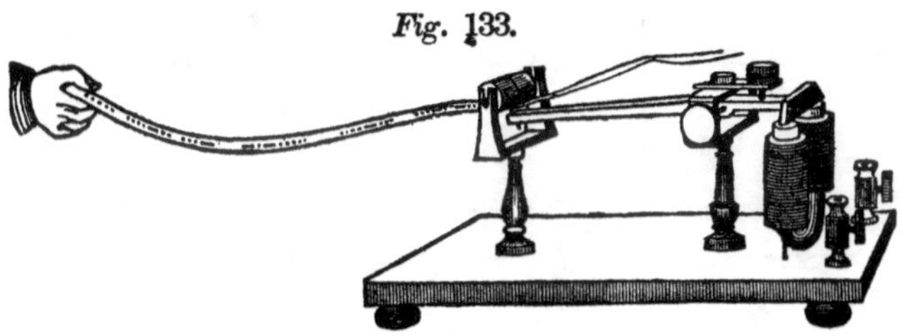

consists of an electro-magnet, armature, and lever, arranged in a similar manner to the magnetometer, (Fig. 131.) At the extremity of the lever is a blunt point, which marks the strip of paper when the electro-magnet is in action. If one wire from the battery is placed in one of the screw-cups, whenever the other wire is touched to the remaining cup, the armature is powerfully attracted by the magnet, and the point on the lever presses the paper into the corresponding groove of the roller, so that lines or dots are made according to the time during which the contact with the battery is maintained, the paper being slowly drawn under the roller.

314. The following table exhibits the signs employed by Professor Morse:—

MORSE'S TELEGRAPHIC ALPHABET.

ALPHABET.		NUMERALS.
	n — -	
a - —	*o* - -	
b — - - -	*p* - - - - -	
c - - -	*q* - - — -	1 - — — -
d — - -	*r* - - -	2 - - — - -
e -	*s* - - -	3 - - - — -
f - — -	*t* —	4 - - - - —
g — — -	*u* - - —	5 — — —
h - - - -	*v* - - - —	6 - - - - - -
i - -	*w* - — —	7 — — - -
j — - — -	*x* - — - -	8 — - - - -
k — - —	*y* - - - -	9 — - - —
l ——	*z* - - - -	0 ——
m — —	*&* - - - -	

It will be observed, that the characters are formed by different combinations of dots, and of short and long lines, with short and long spaces. Between each letter of a word, a short space is used, and long ones between the words themselves. Sentences are separated by still longer spaces.

315. TELEGRAPH, WITH CLOCK-WORK.—In Fig. 134, the telegraph is shown with all the appendages

Fig. 134.

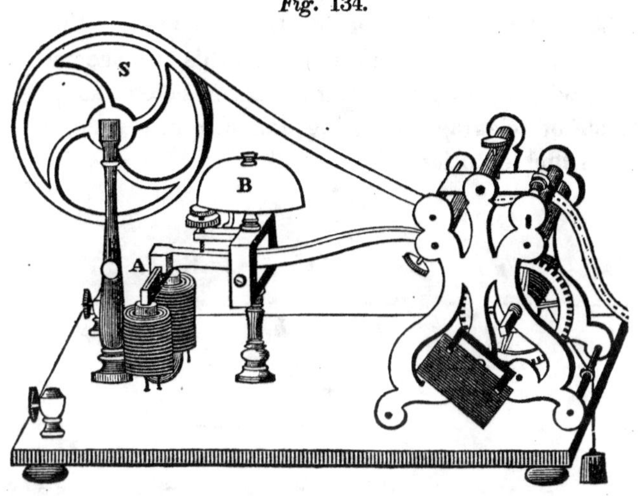

usually employed. The electro-magnet near A, and the lever, are the same as in the instrument last described; connected with them is an apparatus, moved by clock-work, for carrying the paper, which is unwound from a large spool, S, by the rollers between which it passes, and thus drawn slowly over the registering point. The first movement of the lever sets the clock-work in action, and a break and friction-wheel are sometimes added, by which it is stopped shortly after the signals cease to be transmitted. A bell, seen at B, is connected with the lever, so that the first motion communicated by the battery gives a signal to the attendant.

316. The battery used in operating Morse's tele-

graph, is a modified form of Grove's, (§ 45.) Instead of a galvanic series, the magneto-electric machine, to be described hereafter, has been used as a source of electrical power for working the telegraph. The first instance, it is believed, in which this was accomplished with Morse's telegraph, is given below, as an example of telegraphic writing, taken from the scroll used on that occasion:—

·——   · ··   ··   —   —   ·   —
W r i t t e n

—···   ·· ··   —   ····   ·   ·····
b y t h e p

· ·   ·——   ·   · ··   · ·   ·—
o w e r o f

· ·   —·   ·   · ·   ·—·   —··
o n e o f D

·—   ···—   ··   ···   ·   ···
a v i s e s

——   ·—   ——·   —·   ·   —
M a g n e t

· ·   ·   ——   ·   ·· ·   —
o E l e c t

· ··   ··   ·· ·   ——   ·—   ·· ·
r i c M a c

····   ··   —·   ·   ···   —··
h i n e s B

· ·   ···   —   · ·   —·   —··
o s t o n D

·   ·· ·   · ·   —··—   ·——
e c r 9 1

—····   ····—   · ··—.
8 4 4

317. Signal Key. — Fig. 135 represents the instrument usually employed to make the various contacts, differing in succession and length, by which the system of lines and dots, representing the various letters, are produced at the other end of the telegraph. The fingers are shown resting upon a knob, attached to a metallic spring, by the depression of which contact is made with a metallic conductor on the baseboard, connected with a screw-cup at one extremity of the instrument; the other screw-cup is connected with the spring. One battery wire passes into the first screw-cup. The other screw-cup receives one of the wires of the telegraph, which proceeds to the registering apparatus at the other station. By the other telegraph wire, the remaining extremities of the battery and of the registering apparatus are directly connected. The circuit is therefore completed by depressing the key, and is immediately broken by the action of the spring when the fingers are removed.

Fig. 135.

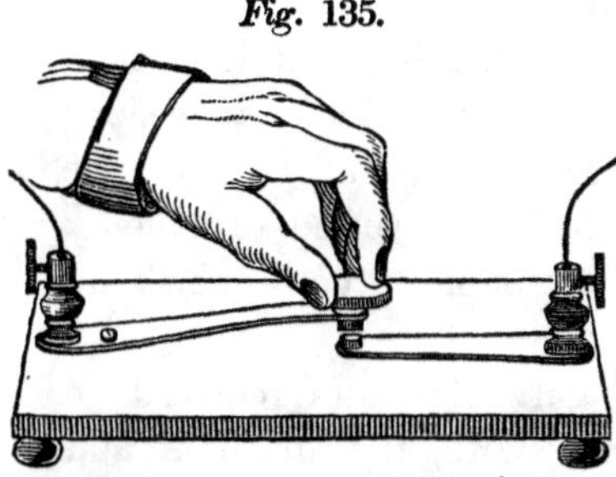

318. Instead of using the second wire directly connecting the battery and register, the earth is sometimes employed as a conductor, by connecting the pole of the battery and register with a large metallic plate, sunk in the ground at each terminus of the telegraph. In this case, the number of wires needed is reduced to one.

319. The wires are well insulated, and are carried from station to station along high posts, about 300 feet apart. On some of the lines they are of copper; on others, of iron. An iron wire is much inferior in conducting power to a copper one of the same size; but the greater cheapness admits of the use of much thicker wires. These are found to convey the current well, and are less liable to be broken than the copper wires.

320. The number of galvanic pairs required to work the telegraph varies with the distance and consequent length of the wires. From twelve to thirty or fifty have been employed upon different lines. The largest number would still be inadequate to give sufficient power to the electro-magnet at very long distances, were it not for the invention of the *receiving magnet*, by Professor Morse. An electro-magnet of the most sensitive construction is placed at the end of the telegraphic line, which moves an armature by the galvanic power, sufficiently only to make contact between a platinum point and surface, which are in the circuit of another small battery, immediately on the spot. This at once works the registering apparatus with great force, its current having to traverse only a few feet of wire in its passage to the registering electro-magnet.

321. Axial Receiving Magnet.—In Fig. 136 is represented a receiving magnet of a new construction. The movement is produced by the axial force, an iron bar of the U form being drawn within a double helix, as described in § 289. The bent bar

is horizontal, but is so suspended that its own weight withdraws it from the coils when the current is interrupted. In the cut, the instrument is shown in connection with a local battery for igniting an explosive charge in blasting rocks, or for submarine or other explosions. The loss of power by con-

Fig. 136.

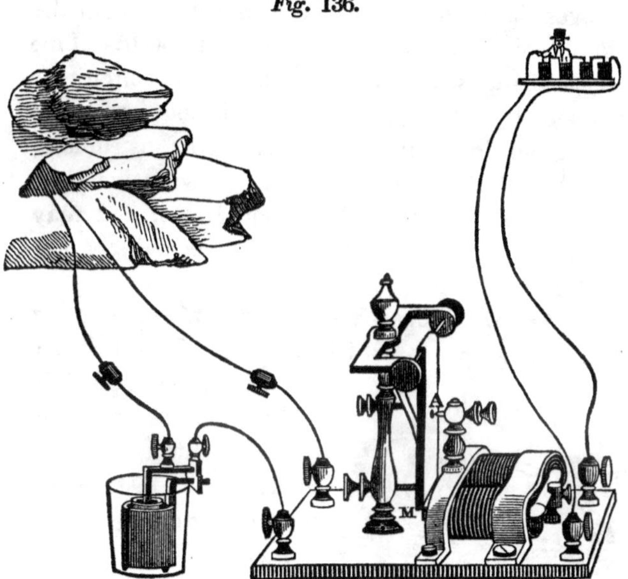

duction, renders it difficult, with a moderate sized battery, to explode the charge by heating a wire at a considerable distance. In the arrangement here described, the receiving apparatus is used to complete the circuit of a small battery in a protected position near the place where the explosion is to occur.

322. In the figure, the coils are placed horizontally, and the U-shaped iron bar, which moves in connection with a vertical bar, suspended by an axis above, enters them at M. The influence of the helices is aided by an armature, as in Fig. 121. The connections of the single galvanic pair, seen in the foreground, are so made that its circuit is completed whenever the vertical bar comes in contact with the platinum point, A, immediately in front of it. This takes place as soon as the long battery circuit is completed by the operator at a distance, and the axial magnet drawn into the coils. This instrument, even without the armature, is more sensitive than the receiving magnet of Professor Morse, and may be used with some advantage as a receiving apparatus for the telegraph.

323. Axial Telegraph. — Fig. 137 represents a form of telegraph in which the axial force is used, motion being obtained in the manner described in § 289. This is introduced as a new instrument, and its originality is here claimed. Two upright coils, C, are supported a short distance above the baseboard. Entering these from below is a U-shaped rod of soft iron, fitted to be drawn up into the coils, under the influence of the galvanic current. When not thus drawn up, it rests on a spring, shown in the cut, by which the instrument is rendered more

Fig. 137.

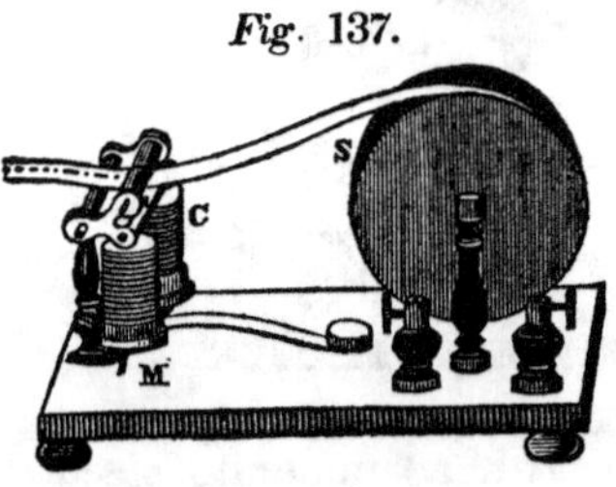

sensitive. A grooved roller, for carrying the paper, is seen above the coils. To the iron bar is attached a blunt point, so as to project above its poles, in the centre, and in opposition to the groove on the roller. Contact being made with the battery, the soft iron rises up within the coils, and marks the paper. On the whole, this form of telegraph is more sensitive and efficient than the electro-magnetic, and less liable to derangement; from the absence of the lever, it is also more compact. A small armature may be placed across the coils above, as described in § 290, by which the mark is made with still greater force; but in this case, the motion is partly electro-magnetic.

324. Axial Telegraph, with Clock-work. — In Fig. 138, the axial telegraph is represented, with

Fig. 138.

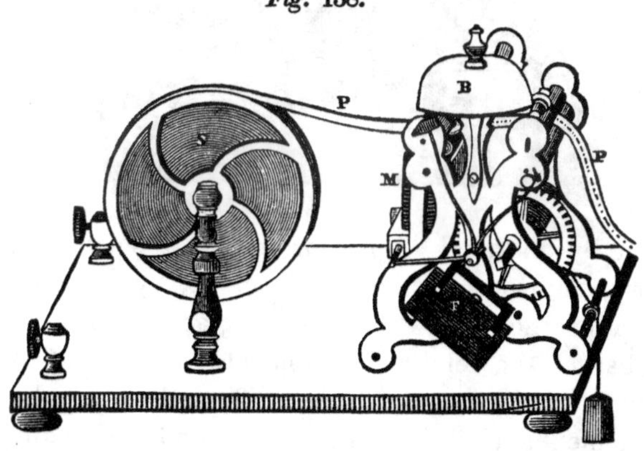

clock-work annexed, for carrying the paper, in the manner already described in connection with the electro-magnetic telegraph. The coils of the axial apparatus are partly seen at M. The large scroll of paper is shown at S, with the strip, P P, proceeding from it. The arrangement for giving the alarm by the bell, B, is also annexed.

325. Axial Telegraph, with Engine. — A different arrangement of the telegraph is shown in Fig. 139, where the machinery for carrying the paper is

Fig. 139.

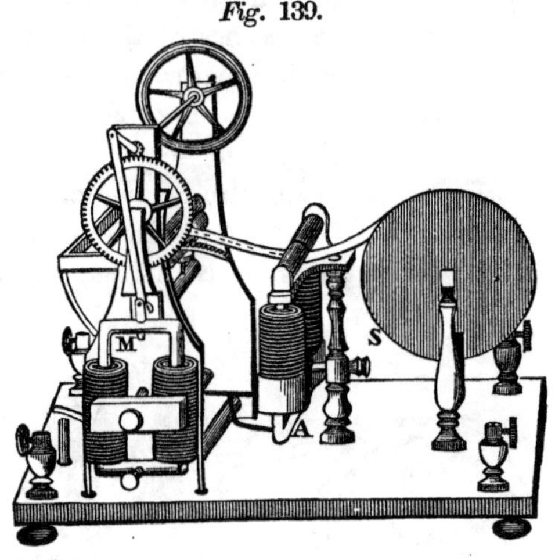

moved by an axial engine, driven by a small local battery, and set in action, if desirable, by the completion of the circuit of the telegraph. The soft iron bar of the telegraphic apparatus is seen at A, and that of the engine at M. There is a peculiar arrangement added for commencing and suspending the action of the engine when the telegraph begins and ceases to work.

326. The deflection of the gold leaf galvanoscope,

(Fig. 63), has recently been proposed as a means of telegraphic indication. The extreme delicacy of this instrument enables it to give a result with a single pair of plates through great lengths of wire. There are no means, however, of recording its action, which would also be seriously interfered with by the influence of atmospheric electricity on the conducting wire. The transmission of the battery current through the telegraphic wire, however long, is practically instantaneous. Experiments which have been made on the velocity of electricity, appear to indicate that it is considerably greater than that of light, which travels nearly 200,000 miles in a second of time.

327. Reciprocating Armature Engine.—In this instrument, contrived by Dr. Page, two electro-magnets of the U form, represented at M M (Fig. 140),

Fig. 140.

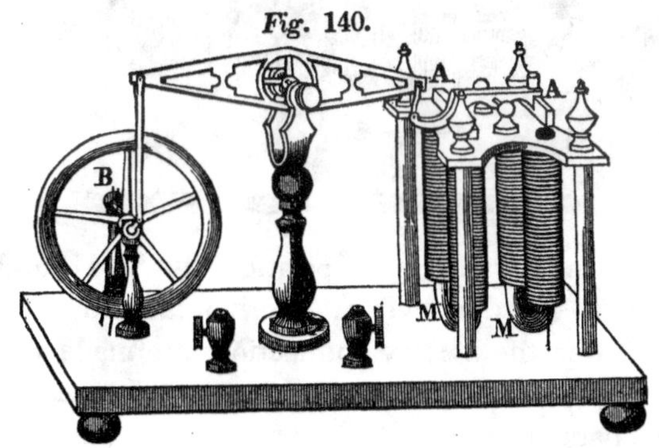

are firmly secured in a vertical position, the four poles appearing just above a small wooden table. The two armatures, A A, connected by a brass bar, move upon a horizontal axis in such a manner that while one is approaching the poles of the magnet over which it is placed, the other is receding from those of the other magnet. The brass bar is connected with one extremity of a horizontal beam, the other end of which communicates motion by means of a crank to a fly-wheel. On the axis of the fly-wheel at B is the break-piece. Each magnet being charged in succession, the armatures are attracted alternately, communicating a rapid reciprocating motion to the beam, and consequently a rotatory one to the fly-wheel.

328. Horizontal Reciprocating Engine. — Two electro-magnets of the U form, M M (Fig. 141), are

Fig. 141.

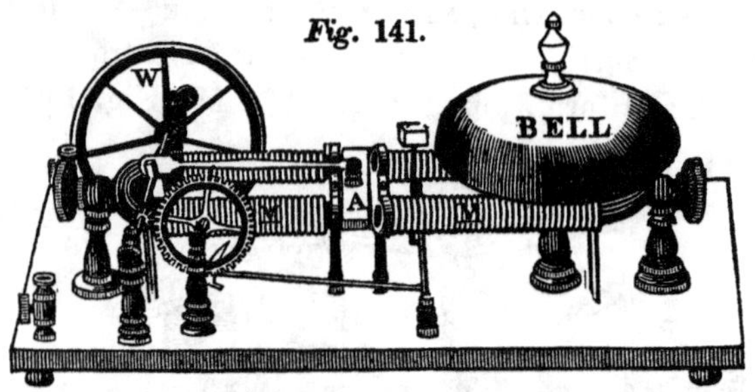

supported in a horizontal position, with a single armature A fitted to vibrate horizontally between them. When the battery connections are made with the screw-cups on the baseboard, one of the magnets will be charged, provided the break-piece is in such a position with regard to the springs as to complete the circuit. The armature will now be attracted towards the charged poles. Just before it reaches them, the movement of the break-piece interrupts the current in the magnet, destroying its polarity, and then causes the current to be transmitted through the opposite one; this becomes charged in its turn, and attracts the iron bar, A, which imparts motion to the balance-wheel, W. By means of clock-work, the rotation of the wheel causes a hammer to strike the bell placed over one of the electro-magnets.

329. Revolving Armature.—In this instrument, invented by Dr. Page, a small bar of iron is arranged to revolve horizontally just above the poles of an electro-magnet of the U form, fixed in a vertical position, as seen in Fig. 142, where A is the iron bar, and M the electro-magnet. To the axis of motion of the bar is affixed a break-piece, made by filing away two opposite sides of a small solid cylinder of silver. Upon the narrow prominent portions thus left, play two silver springs, shown at W in the cut, opposite to each other. One of these springs is connected with a screw-cup on the stand; the other communicates with one extremity of the wire enveloping the electro-magnet, the other end of this wire being fixed to a second cup on the baseboard. The break-piece is so arranged as to release the springs from their bearing just as the

Fig. 142.

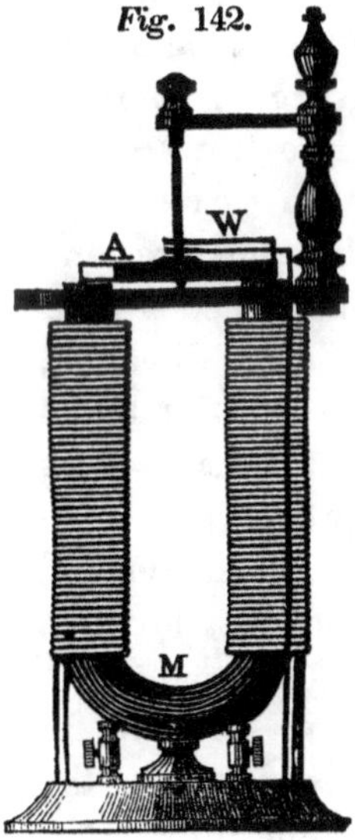

armature passes over the poles; and to restore them to it again when it has moved on somewhat more than a quarter of a circle, so as to be a little inclined from a position at right angles to the plane of the magnet.

330. On placing the bar in this position, and connecting the cups on the stand with a battery, the electro-magnet becomes charged, and consequently attracts the armature towards its poles: as soon as it reaches their plane, the springs leave the projecting parts of the break-piece, and the current is cut off. The polarity of the magnet is now destroyed, and it ceases to attract the armature, which moves on by the momentum it has acquired, until it passes a little beyond a position at right angles to the plane of the magnet. At this point, the springs again come in contact with the break-piece, and the flow of the current is renewed. The attraction now exerted by the poles gives a new impulse to the armature, and the circuit being again broken when it reaches their plane, it continues its motion in the same direction, revolving with great speed. Care should be taken that the springs are in such a state of tension as to open and close the circuit at the proper points, as indicated in the above description. The motion of the bar is not reversed by changing the direction of the current.

331. Revolving Armature Engine. — In this instrument there are several armatures fixed on the circumference of a vertical wheel, parallel to its axis. In Fig. 143 three are represented, each of them

marked A. On the poles of the electro-magnet, M, is secured a brass plate, from which rise two pillars to support the axis of the wheel: as the wheel turns, the iron bars pass in succession over the poles with their extremities very near to them. At B, on the shaft of the wheel, but not insulated from it, is the break-piece, consisting of a small metallic disc, from which project, in a lateral direction, several pins, equal in number to the iron bars; or the disc may be furnished with a corresponding number of teeth. A silver spring connected with one end of the wire surrounding the electro-magnet, plays upon these pins or teeth; the other end of this wire is soldered to the iron of the magnet, which brings it into metallic communication with the shaft by means of the brass plate and pillars. Or the wire may be terminated by a second spring pressing upon a cylindrical part of the axis.

Fig. 143.

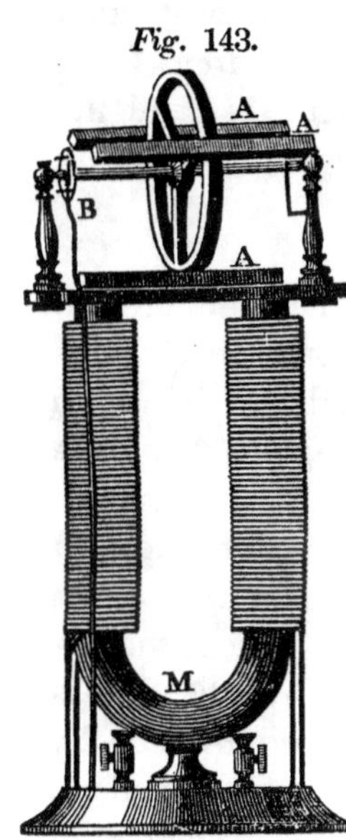

332. The break-piece is arranged in such a manner, that the electro-magnet is charged when any one of the iron bars is brought near it by the motion of the wheel. This bar is then attracted towards the poles; when it arrives at the plane of the magnet, the current is cut off, in consequence of the corresponding pin or tooth releasing the silver spring

from its bearing. The armature being no longer attracted, the wheel moves on by its momentum until the next bar comes into the same position, causing the magnet to be recharged; this is then attracted in its turn, and passes on like the preceding one.

333. The spring playing on the break-piece must be so disposed that the circuit shall be broken when each bar reaches the poles, and not be renewed again until it has passed to a greater distance from them than that between the next succeeding bar and the poles, or it will be attracted back again, preventing the continuance of the motion.

334. Vibrating Armature Engine. — Fig. 144 represents an instrument in which the armature, A, is made to vibrate backwards and forwards above the poles of an electro-magnet, M. The armature, when on one side, is attracted by the poles until it arrives directly over them. As it reaches this position, the circuit is broken by the break-piece, and the bar moves on by the momentum it has acquired, for some distance. The circuit is then renewed, and it moves back towards the poles. The arrangement of the break-piece is such, that the current circulates while the armature is approaching the poles on

Fig. 144.

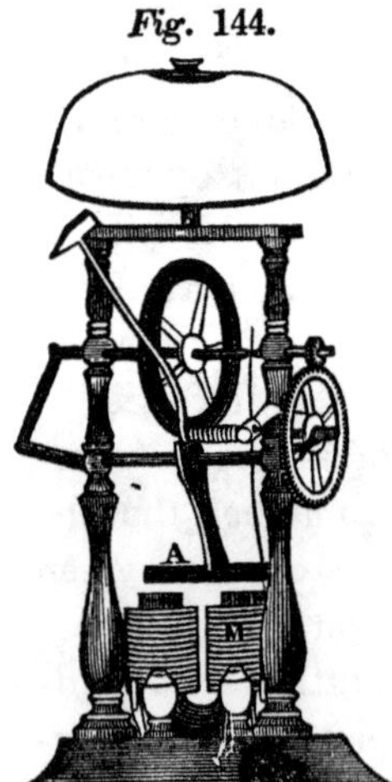

either side, and is interrupted while it is receding from them. A continuous vibration is thus occasioned, which imparts motion to a fly-wheel, and to a hammer, which strikes the bell seen in the cut.

335. Fig. 145 represents another form of the instrument, in which the axis of motion of the armature is below the magnet. The moving bar exerts its greatest force when passing the poles, and at the moment when the crank is in the best position for driving the fly-wheel.

Fig. 145.

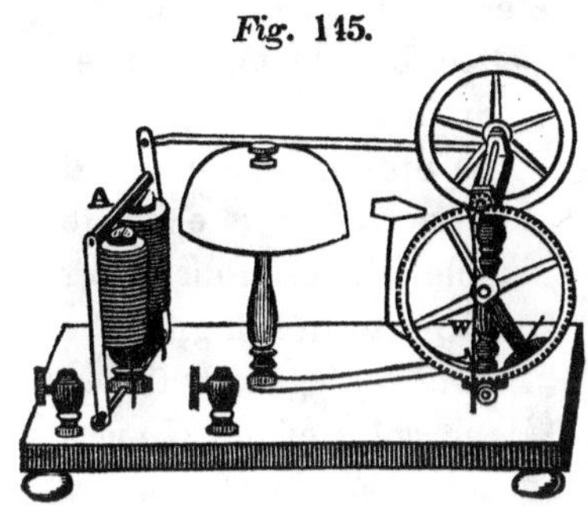

336. Revolving Electro-Magnet.—In the instrument represented in Fig. 146, a steel U-magnet is fixed in a vertical position, and a small straight bar of soft iron, enclosed in a helix, is so arranged as to revolve between its poles. The two extremities of the insulated wire surrounding this electro-magnet, are connected respectively with the segments of a pole-changer (§ 191), on the shaft. The silver springs, which press upon the pole-changer, are attached to two stout brass wires, passing through the brass arch surmounting the U-magnet, but insulated from it by the intervention of ivory or horn; each of these wires supports a brass cup for connection with the battery. These springs must be so placed with regard to the segments, that the poles of the

revolving bar shall be reversed at the moment when it is passing the poles of the fixed magnet.

337. On making connection with the battery, when the bar is at right angles to the plane of the magnet, it immediately acquires a strong polarity. Its north pole is then attracted by the south pole of the steel U-magnet and repelled by the north pole. The south pole of the bar, on the contrary, is repelled by the similar pole of the upright magnet, and attracted by its opposite pole. These four forces conspire in bringing the electro-magnet between the poles of the U-magnet. When it reaches this position, each segment of the pole-changer leaves the spring with which it was in contact, and passes to the other. As the bar is moving past the poles by the momentum it has gained, its magnetism is destroyed for a moment, and immediately restored in the opposite direction. Each pole of the bar is now repelled by that pole of the permanent magnet which it has just passed, and attracted by the opposite one;

Fig. 146.

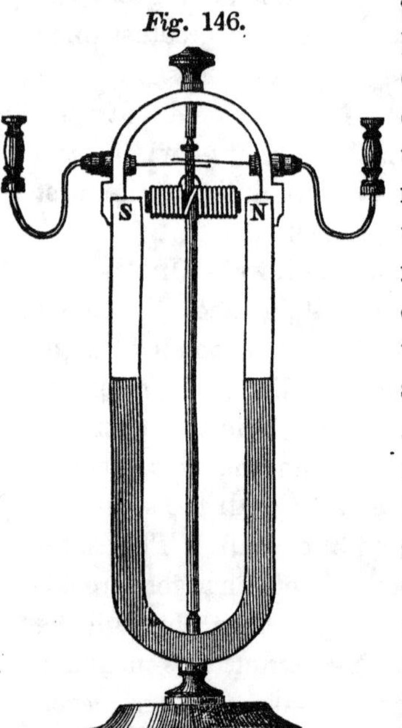

it consequently moves on, the polarity being reversed twice in each revolution.

338. Instead of using a pole-changer, the poles of the electro-magnet are reversed in the following manner in the instrument known as RITCHIE'S REVOLVING MAGNET. The ends of the wire surrounding the revolving bar dip into mercury contained in a circular cistern of ivory, fixed between the poles of the U-magnet below the bar. This cistern is divided into two separate cells by low partitions of ivory, so arranged that, when the electro-magnet is passing between the poles of the steel magnet, the ends of the wire may be moving across the partitions and just above them. On supplying the cells with a proper quantity of mercury, its surface will be found to curve downwards on every side towards the ivory, so that its general level will be higher than the partitions; thus allowing the extremities of the wire to be immersed in it, except when passing across them. A wire connected with a brass cup, for making communication with the battery, projects into the mercury in each compartment of the basin. At the moment when the wires quit the mercury to pass across the partitions, a spark is seen. When the machine is put in motion in a dark room, these sparks give rise to an optical illusion of the same character as that mentioned in the description of the Revolving Spur-Wheel, causing the bar to appear at rest in the position it is in when the sparks are emitted.

339. The revolution is more rapid where the pole-changer is used, than in Ritchie's instrument,

and may be made still more so, by employing a U-shaped electro-magnet in place of the stationary steel magnet. In this case, the rotation is not reversed by changing the direction of the current, as it is when a steel magnet is used, since the poles of both electro-magnets are reversed at the same time, and their relative polarity remains the same.

340. Revolving Bell Engine. — This instrument, represented in Fig. 147, is similar in principle to the one last figured, the U-magnet, however, being inverted, so that the revolving electro-magnet is near the baseboard; the pole-changer is on the axis below it. There is, in addition, an arrangement for striking a bell, which is fixed above the magnet. To the axis of the revolving bar is attached an endless screw; this screw acts on a toothed wheel, which is provided with a pin projecting laterally, for the purpose of moving the hammer. As the wheel turns, the pin presses upon the handle of the hammer, raising it from the bell until it is released by the pin at a certain point of the revolution; when a spiral spring, fixed to the handle, impels the hammer against the bell.

Fig. 147.

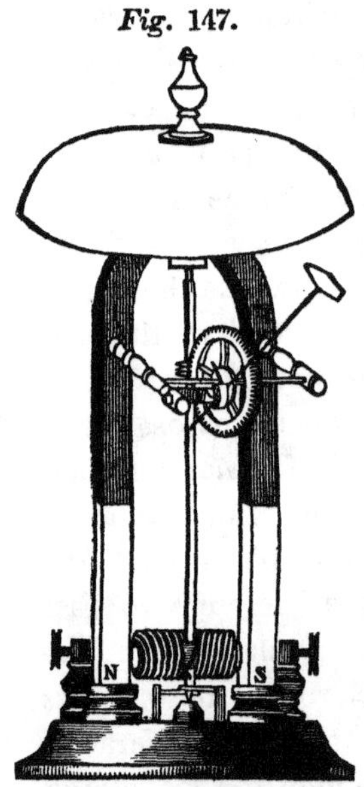

341. If the wheel has 100 teeth, as in the cut, the electro-magnet must revolve 100 times in order to produce one revolution of the wheel, and consequently one stroke upon the bell. The velocity of the rotating bar is measured by counting the number of strokes in a given time; it may make 100 or more revolutions in a second. In order that the motion of the wheel may raise the hammer, it is necessary to transmit the battery current so that the bar shall rotate in the proper direction.

342. Registering Revolving Magnet. — Fig. 148 represents an instrument in which the number of revolutions of the electro-magnet in a given time is registered by means of clock-work connected with the shaft. On the dial-plate are three pointers, which mark the revolutions up to 1000. The time occupied by any known number of revolutions at once enables us to ascertain the speed of the rotating bar. Though very great, it is necessarily somewhat less than when the bar carries no machinery.

Fig. 148.

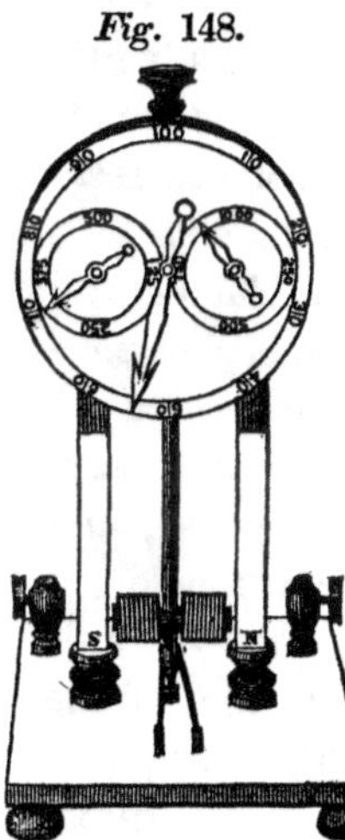

343. Electro-Magnet revolving by the Earth's Action. — As the earth itself exhibits magnetic polarity, an electro-magnet may be made to revolve by its influence; though, in consequence of the feebleness of the action, the instrument must be

constructed with some delicacy. A small electro-magnet, (Fig. 149,) is so supported as to have freedom of motion in a vertical plane, like the dipping needle, a pole-changer being secured on its axis of motion. The springs which press upon the pole-changer should be disposed in such a manner that the polarity of the bar may be reversed when in the course of its revolution it reaches the line of the *dip*. In high latitudes, it will be sufficient to arrange the pole-changer so as to reverse the poles when the bar becomes vertical. The shaft on which the bar revolves, rests on friction rollers.

Fig. 149.

344. On placing the electro-magnet horizontally in the magnetic meridian, that is to say, with its extremities directed north and south, and transmitting the battery current, its north pole (in this hemisphere) immediately inclines downwards towards the earth, in the same manner as that of the dipping needle. The momentum thus acquired carries it past the line of the dip into a vertical direction. As soon as it reaches this position, the poles are reversed, and it continues to move on in the same direction as long as the battery connections are maintained, revolving with a moderate velocity.

345. In Fig. 150 is represented an electro-magnet,

so supported as to revolve by the earth's action in a horizontal plane, instead of a vertical one. In this case, the instrument must be placed in such a position that the polarity shall be reversed when the revolving magnet assumes the direction of the compass-needle, pointing north and south.

Fig. 150.

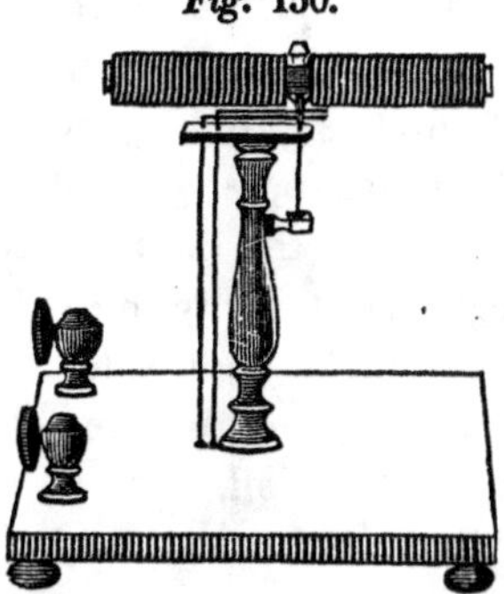

346. When a steel magnet is placed in a proper position near the revolving bar, in either of the instruments whose motion depends on the earth's influence, it rotates with much greater speed than by the action of terrestrial magnetism alone. Its motion may be reversed, notwithstanding the opposing influence of the earth, by disposing the permanent magnet in a suitable manner.

347. ELECTRO-MAGNET REVOLVING WITHIN A COIL. — Fig. 151 represents an instrument in which a straight electro-magnet revolves within a circular coil. The same current traverses the coil and the wire surrounding the rotating bar, but its direction is changed only in the latter. When the circuit is completed, the revolving bar moves so as to bring its poles into a direction corresponding with those of the coil. This motion depends upon the same cause as that of a galvanometer needle, (§ 158.) As the poles of the coil are situated in the line of its axis, the springs pressing on the segments of the pole-changer

are so arranged as to reverse the current when the electro-magnet is at right angles with the coil. The north pole of the magnet places itself within the north pole of the coil, and not, as might at first sight be expected, within its south pole. This apparent anomaly is due to the circumstance of the magnet not being presented to the coil from the outside, but lying within, with its centre coinciding with that of the coil. If a magnetic needle is brought up to a helix, its north pole is attracted towards the south pole of the helix, (see § 268), and passes within it, until its centre lies in the centre of the coil. In this position, its poles lie in the same direction as those of the electro-magnet described above.

Fig. 151.

348. Double Revolving Magnet. — In this instrument, represented in Fig. 152, there are two electro-magnets of the U form and of the same size, arranged to revolve on vertical axes. At s, on the shaft of the lower magnet, is a silver ferrule, on which presses a wire connected with one of the screw-cups. On the shaft of the upper magnet is a break-piece. When

Fig. 152.

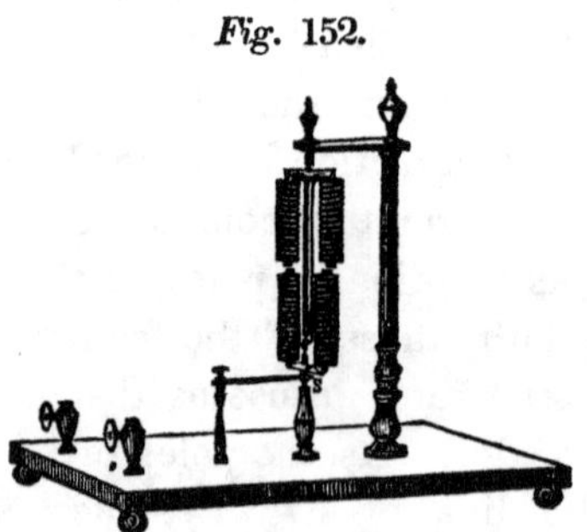

the current passes, both magnets become charged, and, their opposite poles attracting each other, they move in opposite directions until they come into the same line. At this moment the circuit is interrupted by the break-piece, and the poles move past each other by their acquired momentum. The circuit is then renewed, and the magnets continue to revolve in opposite directions. Since the shaft of the lower magnet sustains the weight of both, the lower one moves with less speed than the other, in consequence of the increased friction. Reversal of the poles by a pole-changer is not resorted to, on account of the feeble repulsion between electro-magnets.

349. Revolving Wheels, with Electro-Magnet. — In the instrument represented in Fig. 153, motion is obtained on the same principle as in the Revolving Spur-Wheel (§ 180), an electro-magnet taking the place of the steel magnet. There are two wheels; of these, the upper and larger one is partly supported by springs, its circumference resting on that of the smaller wheel.

Fig. 153.

The current traverses in succession the coil of the electro-magnet and the wheels. It passes from the axis of the smaller wheel to that part of its circumference which touches the circumference of the other, and thence to the axis of the larger one. The lower wheel is not merely carried round by the other, but those parts of both which are conveying the current tend to move away from between the poles in the same direction, causing the two wheels to revolve in opposite directions.

350. Revolving Wheels. — A better form is shown in Fig. 154, in which the wheels move between the poles of a steel U-magnet, whose legs are brought very near together. In these instruments no mercury is used. Fig. 154 is similar to the Revolving Disc (§ 184), and would have been described in connection with that, had not that part of the volume gone to press before this instrument was contrived.

Fig. 154.

351. Electro-Magnet, revolving on its Axis. — The instrument represented in Fig. 155, is similar to the one described in § 167, except that the revolving bar is of iron, enclosed in a helix which rotates with it. The battery current traverses the helix and one half of the bar. As the bar becomes an electro-magnet, it revolves like the one represented in Fig. 57, except that the direction of the rotation is not changed by reversing the current, since the poles are at the same time reversed. From the power of the magnet, the motion is very rapid.

The revolution of either an electro-magnet or a steel magnet on its axis is properly classed with that of a conductor around a magnet. When the conductor and magnet are fastened together, the latter is carried round by the movement of the former, as stated in § 171. Instead of fastening a conducting wire to the magnet, the current may be transmitted through the magnet itself, with the same result.

Fig. 155.

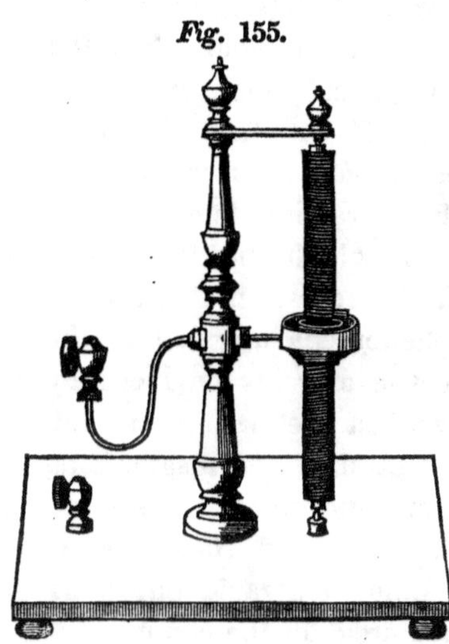

352. Electro-Magnetism as a Motive Power. — The great velocity of motion, and the strong attractive force exhibited by many of the small electro-magnetic instruments, naturally suggested the application of this power to the purposes of the arts as a mechanical agent; and numerous experiments have been made with this view, but hitherto without success. Professor Henry was the contriver of the first instrument whose motion depended upon magnetic attraction and repulsion. In his little machine, an electro-magnet, whose polarity was alternately reversed, was made to vibrate above the north poles of two straight steel magnets. He, however, made

no attempt to apply this power to practical purposes There are obstacles of a purely mechanical character in the way of its employment; these, though important, are not perhaps insurmountable. But the most serious difficulties are those which arise from the nature of the power. The motion of the attracting poles of two electro-magnets towards each other, actually lessens the attractive force in proportion to the velocity with which they approach; the same thing occurs in the recession of mutually repelling poles. These phenomena are due to the influence of secondary electric currents produced by the motion, which flow against the battery current, and of course partially neutralize its magnetizing power. The secondary currents present a very formidable obstacle, as their opposing influence increases with the size of the machine in a rapid ratio. An appreciable time is also required for the communication and removal of the polarity of large electro-magnets, and the purest iron retains a considerable degree of magnetic power after being highly charged. To the action of these and some other causes is owing the fact, which was early discovered by those engaged in these investigations, that the smallest machines possess by far the greatest proportional power. Some of these difficulties are obviated by the axial motion, recently proposed as a source of mechanical power, by Dr. Page, one of the earliest and most persevering, as well as ingenious experimenters in this department. In his instruments, modified forms of some of which have been described on previous pages, the force by which an iron bar is drawn into a helix is the source of the power. A stroke of some length may thus be obtained, and there are no attracting or repelling poles to produce interfering secondary currents.

III. BY THE INFLUENCE OF THE EARTH.

353. An unmagnetic bar of iron, placed in the direction assumed by the dipping needle (see § 210), acquires temporary magnetism by induction from the earth. That extremity which is directed towards the south magnetic pole (§ 218), receives north polarity, and the other end south polarity. A bar of soft iron held in a horizontal position, especially if directed east and west, attracts indiscriminately either pole of a magnetic needle, as the earth exerts no appreciable inductive action upon it.

354. In Fig. 156, A B represents a bar of iron, presented in a horizontal position to the north pole of a magnetic needle, N S. The pole is now attracted by the bar. Keeping the end B in the same place, raise the end A so as to bring the bar into the position C D. As the bar is raised, the north pole recedes from C, as indicated by the dotted lines in the cut. The strongest action is exerted when the bar is in the line of the dip, or in this latitude nearly vertically over the needle. By carrying the bar below the needle, still preserving the same direction, its upper end will be found to attract N and repel S. These

phenomena are sufficient to show that C D has become a magnet, with C for its north pole. On re-

Fig. 156.

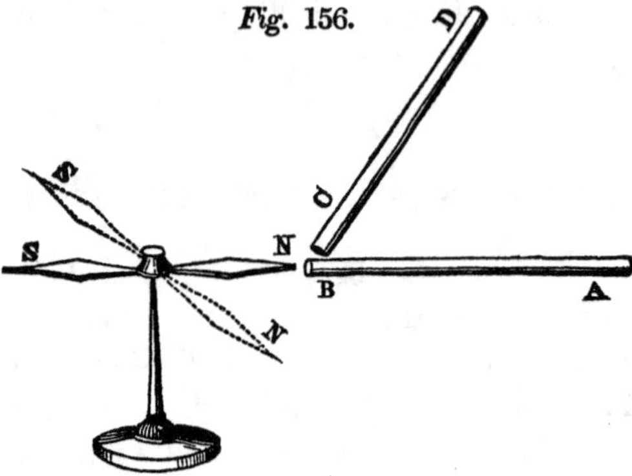

versing the bar, so as to bring the end D downwards, C immediately becomes the south pole: thus the polarity may be changed at pleasure, the induced magnetism being only temporary. If the bar is brought very near to the pole of the needle, the inductive action of the earth will be overpowered by that of the needle, causing attraction to be exhibited in every position.

355. Except in the vicinity of the equator, it is sufficient to hold the bar vertically, as the line of dip approaches the perpendicular in high latitudes. In consequence of this inductive action of the earth, all large bars of iron standing in an upright position, are more or less magnetic, their lower ends, in this hemisphere, being north poles. When they have remained for a long time in this situation, the polarity does not disappear on changing their position.

356. The induction of magnetism by the earth is greatly facilitated by causing a movement among the particles of the bar, as by percussion or twisting. Let a bar of iron be held in the line of the dip, and its lower end brought near the north pole of a magnetic needle. The pole will be repelled, and will come to rest at some distance from the bar, as shown in Fig. 157. Now strike the upper end of the bar with a hammer, and the induced magnetism will be so much increased that the needle will swing round, and its south pole be drawn to the bar, as indicated by the dotted lines in the cut. The polarity thus induced is not reversed by merely inverting the rod, but the aid of percussion will be required, in order to remove or reverse the magnetism. This is especially the case when a steel bar is employed instead of an iron one.

Fig. 157.

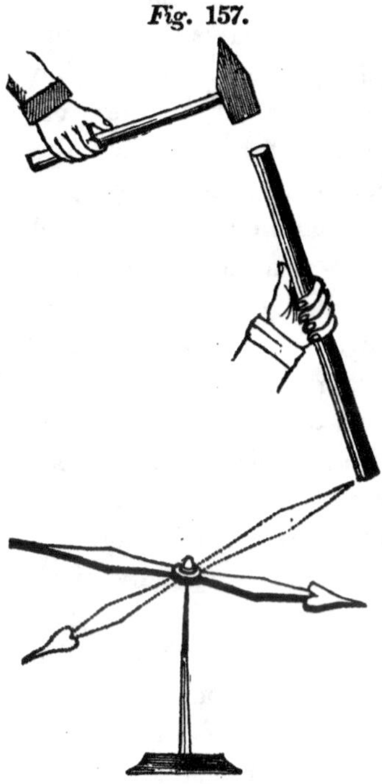

357. Take a piece of iron wire, and, placing it in a vertical position, twist it powerfully. It will now be found to sustain iron filings at its extremities, and

to turn itself north and south, when balanced on a pivot, as shown in Fig. 158. The end which was downwards becomes the north pole.

Fig. 158.

358. The magnetism, in these cases, is not due directly to the percussion or twisting, either of which merely favors the action of the earth. A considerable degree of permanent magnetism may be communicated to a steel bar, by placing it vertically on a large mass of iron, and striking its upper end repeatedly with a hammer; it acquires much greater power if struck while resting on iron than on any other substance.

359. Percussion may be used to facilitate the removal of magnetism. Thus the polarity of a steel bar magnet may be lessened, or even entirely destroyed, by repeated blows of a hammer, while held horizontally east and west. This process is very convenient for removing slight degrees of magnetism from iron or steel bars. In discharging steel magnets by the methods previously described, it is sometimes difficult to effect a complete removal of their magnetism. In such cases, the desired result may be attained, after having nearly discharged them, by following the process now given. Merely falling upon the floor often injures the power of a magnet considerably, in consequence of the vibration excited among the particles of the steel.

MAGNETISM.

III.

INDUCTION OF ELECTRICITY.

I. BY THE INFLUENCE OF A CURRENT OF ELECTRICITY.

360. THAT branch of the science of electricity which treats of the phenomena presented by it when at rest, is termed *Electro-statics:* the branch which relates to electricity in motion, is called *Electro-dynamics.* The phenomena which characterize the latter state are classified by Professor Faraday as follows: "The effects of electricity in motion, or electrical currents, may be considered as, 1st, Evolution of heat; 2d, Magnetism; 3d, Chemical decomposition; 4th, Physiological phenomena; 5th, Spark." Many of the phenomena presented by electricity in motion are closely related to magnetism, and it is usual to treat of them in connection with that subject, as in the present volume, rather than with electricity.

361. Before entering upon the particular subject of the present chapter, that is, the inductive action

of currents, it will be advisable to occupy a few pages with a comparison of some of the phenomena exhibited by electricity in the two states of motion and rest, as induction is exerted in both. The nature of the action is entirely different in the two cases.

362. An electrified body attracts light substances in its neighborhood, having previously induced in their nearest ends the opposite electricity to its own; and on their approach communicates to them a part of its charge, when, if insulated, they are instantly repelled by it. A wire conveying a current exerts no such influence upon light bodies, although placed in the immediate vicinity.

363. In the case of electricity at rest, two bodies, charged either positively or negatively, repel each other; while if one is charged with positive and the other with negative electricity, they exert a mutual attraction. Electrical currents, on the contrary, *attract* each other when flowing parallel in the same direction, and *repel* each other when flowing in opposite directions. The result is the same whether two different currents or two portions of one current are experimented on.

364. Attracting and Repelling Wires. — The instrument represented in Fig. 159 is designed to exhibit the attractions and repulsions of currents. Two wooden troughs for containing mercury are supported opposite to one another, each being divided into two oblong cells by a partition in the middle. Each of the four portions of mercury thus insulated,

is connected by means of a wire projecting into the cell, with one of the screw-cups fixed at the ends of the troughs. The points of two rectangular wires, A and B, rest in the opposite compartments of the troughs. This mode of support allows the wires to be placed nearer to or farther from each other, at pleasure, still remaining parallel. These wires are balanced by two brass balls, *b b*, attached to them below, which are capable of being raised or depressed by means of a screw cut on the wire; they may thus be so adjusted that the wires will be moved from their vertical position by a very slight force, their upper portions rocking towards or away from each other, without requiring any motion of the points of support.

Fig. 159.

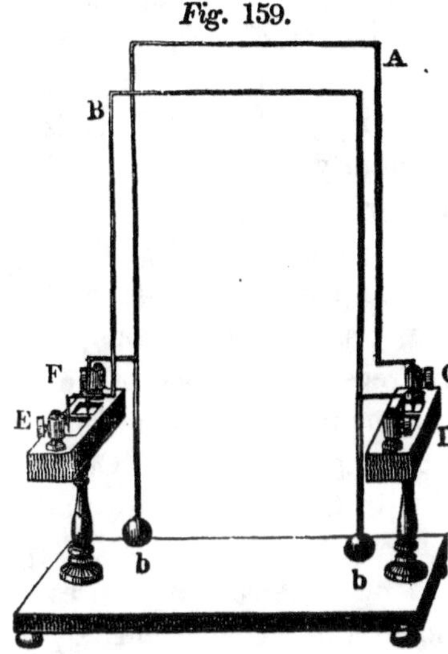

365. Cups C and E, being united by a copper wire, connect cups D and F with the galvanic battery. The current will now traverse A and B in succession, flowing in the same direction in both, and they will be seen to incline towards each other. The motion is slight, but may be made considerable by breaking and renewing the circuit in correspond-

ence with their oscillations. The same effect is produced by uniting D with F, and connecting C and E with the battery. If a powerful current is employed, the wires still attract each other when separated to a considerable distance, by moving the points which rest in the mercury to the farther ends of the cells; with a feeble battery, the wires should be placed near to one another.

366. Let the cup C now be united with D, and E and F be connected with the battery. This will cause the current to flow in opposite directions in the two wires, and they will recede from each other; the extent of the motion may be increased, as before, by alternately opening and closing the circuit. Cups C and D may be connected with the battery with the same result, E and F being united by a wire.

367. The current, instead of traversing the wires in succession, may be divided into two portions by uniting C with D, and E with F, by two wires, and then connecting the battery either with C and F or D and E. In this case, the two portions of the current flow in the same direction in A and B, causing them to attract each other. By uniting C with E, and D with F, the currents in A and B flow in contrary directions, and the wires exhibit a mutual repulsion. The movements produced by a divided current are feebler than when it traverses the wires in succession, unless the battery employed is so powerful that one of the wires singly is not able to convey the whole of the electricity supplied by it.

368. These attractions and repulsions are some-

times called *magnetic*, the two currents, when flowing side by side, acting upon each other like two magnets presented end to end. In fact, if two short pieces of iron wire be suspended end to end, and at right angles to the conducting wires, the magnetism induced in them by the currents (see § 254) will cause them to exhibit similar attractions and repulsions to those of the wires themselves. It is, however, preferable to regard this peculiar action as a primary one; it being highly probable that the polarity of even a steel magnet is due to electric currents circulating within its substance. The mutual actions of two magnets, or of a magnet and a current, would thus be secondary effects, depending upon the attractions and repulsions just described.

369. It is not essential that the current should traverse metallic wires in order to produce these effects. Two streams of machine electricity flowing through a vacuum, or even through the air, exhibit the phenomena very satisfactorily. The attraction of currents moving in the same direction may be shown by the following arrangement: Connect the inner coatings of two Leyden jars with either the positive or negative conductor of a common electric machine, their outer coatings being insulated sufficiently from each other to prevent the passage of a spark between them when the jars are discharged in the mode about to be described. With the exterior coating of each jar is connected a wire having one end free. These ends are left free for the purpose of being placed on a card over which the charge is

to be passed. The common enamelled cards should be used, as they receive a dark-colored and permanent mark from the passage of the spark over their surface. A third wire, attached to the discharging rod, is also to rest on the card, at such a distance from the other two wires that the sparks from the jars may be able to pass. The ends of the wires proceeding from the outsides of the jars should be placed a quarter or a half of an inch apart, and nearer to one another than to the third wire, which is to be equally distant from both, so that two straight lines drawn from them to the third wire, would form the letter V. The jars being charged (during which process the exterior coatings should, of course, be uninsulated), arrange the points as directed, and bring up the ball of the discharging rod to the conductor. The inner coatings being connected, and the outer ones insulated, the current is obliged to divide into two portions as it proceeds from the point attached to the discharger to those in connection with the outsides of the jars. The two sparks will thus pass simultaneously over the surface of the card, and, were they not affected by each other, would leave a mark in the shape of the letter V. It will be found, on the contrary, that the track left on the card will be more or less in the form of the letter Y, the two currents coalescing in their passage over its surface. The result will be the same, whether the jars are charged positively or negatively on the inside. If the wire connected with the discharger is placed under the card while the others are on the upper side, the card will be perforated in one or more places by the passage of the electricity.

370. The experiment may be varied, by connecting with the discharging rod a wire whose ends may both rest on the card at the same distance from each other as that between the two wires attached to the exterior coatings of the jars. The two sets of points being arranged parallel to each other, and their distances properly adjusted, the two currents will remain separate during the whole of their passage over the card; and it will be seen by the marks which they leave, that, instead of proceeding in straight and parallel lines, they form curves whose convexity is turned towards each other. The curvature of the lines is greater in proportion to their proximity; but if the points are placed too near together, both currents flow in one track, not separating until they reach one of the wires connected with the outside of the jars. The resistance of the air and other causes often occasion a stream of electricity to follow a very crooked path in passing over a card. Hence the lines traced by the two currents in these experiments may be irregular, though the tendency to converge is perfectly evident.

371. Contracting Helix. — The mutual attraction between different portions of the same current moving in the same manner, may be rendered evident by the instrument represented in Fig. 160. A wire, coiled into a loose helix, is supported in a vertical position by a brass pillar connected with one of the screw-cups on the base board. The lower

end of the helix dips into mercury contained in a glass cup with a metal bottom, by means of which the mercury is brought into connection with the other screw-cup. When the battery is applied, the portions of the current traversing the different turns of the helix of course flow parallel to each other, and in the same direction. Their mutual attractions cause the turns of the helix to approach each other, shortening it sufficiently to lift the end of the wire out of the mercury. This interrupts the current, and the helix is lengthened again by the elasticity of the wire composing it, producing continued vibration. A spark is seen in the glass cup at each rupture of contact. It is necessary to adjust the quantity of mercury so that the wire may be raised out of it when the coil shortens itself.

Fig. 160.

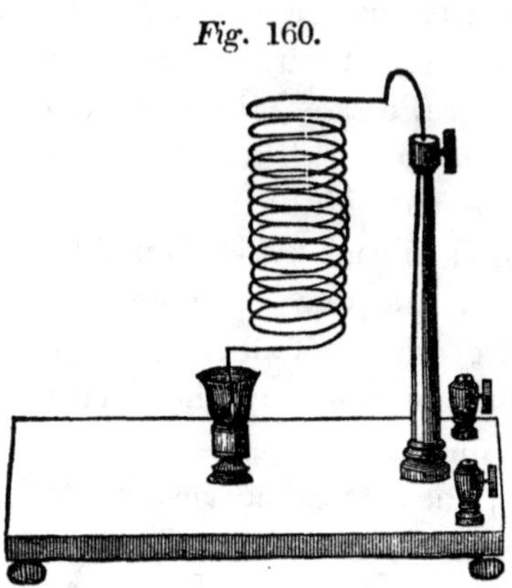

372. A bar magnet passed partly within the helix, as in Fig. 161, causes a more active movement. One pole of a U-magnet, or an iron bar, produces the same effect. If a bar magnet or an iron rod, of the same length as the helix, is passed wholly within, the turns of the helix contract, above and below the centre, while near the centre there is no move-

ment, the influence of the magnet maintaining it in a state of rest. When an iron bar is introduced, it

Fig. 161.

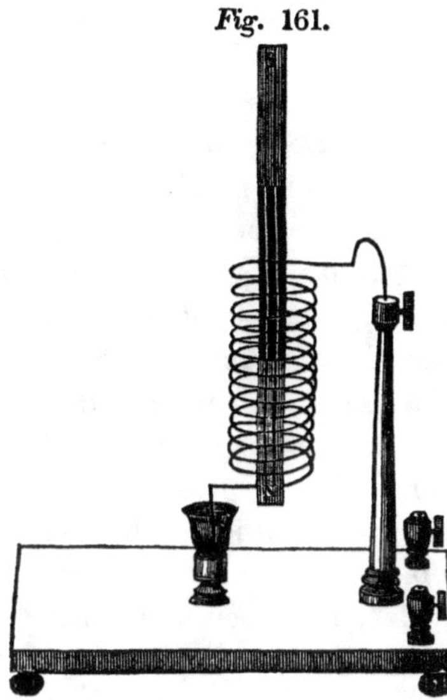

is magnetized by the current in the helix, and exerts the same influence as the steel magnet. The action on each turn of the helix is the same as with De la Rive's Ring, (§ 186.) The direction of the current determines in which direction the steel magnet must be introduced, in order that the helix may contract and break the circuit.

373. Electro-Dynamic Revolving Coil. — The mutually attractive and repulsive action of currents may be made to produce rotations analogous to some of those strictly called electro-magnetic; as in the instrument represented in Fig. 162, which consists of a circular coil of insulated wire, B, fitted to rotate on a vertical axis within a larger one, A, mounted on a brass pillar. The inner coil has a pole-changer fixed to its axis of motion for the purpose of reversing the current twice in each revolution. The current may traverse the two coils in succession, or be divided between them, but its direction must be changed only in the interior one.

374. The inner coil being placed at right angles to the other, and the cups on the stand connected with the galvanic battery, the faces of each coil immediately exhibit north and south polarity, like those of De la Rive's Ring (§ 185); and B is obliged to make a quarter of a revolution in order to bring its north pole within the north pole of A, so as to make their directions correspond. As soon as it reaches this position, the current is reversed by means of the pole-changer, and its south pole now being within the north pole of A, it continues to move on in the same direction. The motion, in this case, depends upon the same principle as those of the wires in the instrument represented in Fig. 159; but it is more convenient to refer them to the polarity exhibited by a current flowing in a circle, as was done in describing the Revolving Coil, (Fig. 69.)

Fig. 162.

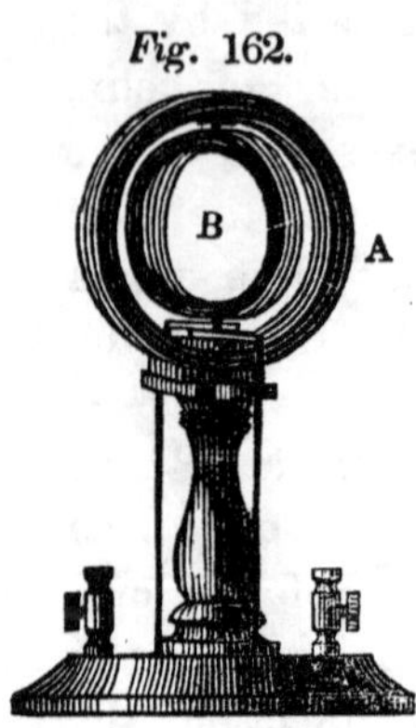

375. It is, however, easy to explain the revolution with direct reference to the mutual action of the currents. As these circulate in the same direction in every convolution of each coil, they may be regarded as two single circular currents. Now, suppose the current in A to be ascending by its left side and descending by its right side. If B is placed at right angles to A, with the current descending in the side towards the spectator, this side will be attracted by

the right side of A, and repelled by the left. The farther side of B, on the contrary, will be repelled by the right side of A, and attracted by its left side. These forces conspire in bringing the two coils into the same direction; when, the current being reversed by the pole-changer, each side of B is repelled by that side of A which is nearest, and the motion is continued in the same direction.

376. Electro-Dynamic Revolving Rectangle. — Rectangular coils may be used in place of circular ones, with the same result, as in the instrument represented in Fig. 163, which is analogous to the one described in § 193. The inner rectangle revolves in the same manner as the coil B in the instrument last described, and the rotation is more rapid than with the circular coil. In consequence of the width of the rectangular coils being less than their height, the sides of the interior one are near those of the other during the whole revolution. This circumstance is the occasion of its greater speed.

Fig. 163.

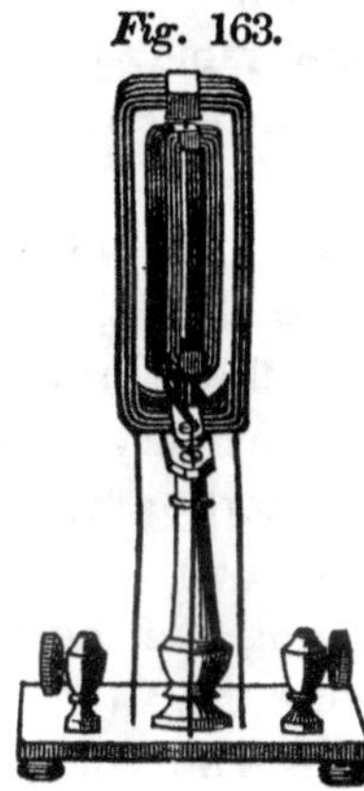

377. Instrument for showing the Revolution of Mercury. — The instrument represented in Fig. 164, is designed to exhibit the revolution of mercury within a helix. The mercury is contained in a brass cup, within which is fixed a small watch crystal. The brass is covered by this, except at the upper edge, where a ring is exposed, which is amalgamated.

This ring is in metallic connection with one of the screw-cups on the base board. Surrounding the mercury cup is a helix, one of whose ends connects with the remaining screw-cup. Its other end is left free, so as to be inserted into the shallow cup. An iron cup, with its rim amalgamated, and a watch crystal secured within it, may be substituted for the brass cup. The movement of the mercury is then more rapid, as the influence of the iron, which becomes an electro-magnet, aids that of the helix.

Fig. 164.

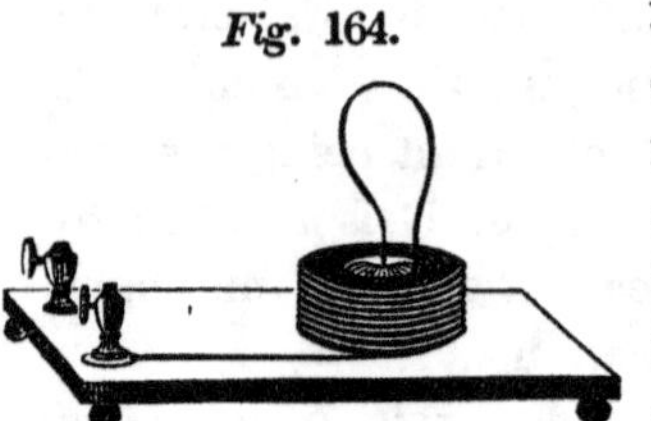

378. Sufficient mercury being introduced, and the battery connections made, insert the free end of the wire into the mercury. The current now flows in a circular direction in the helix, but in the mercury it passes from the end of the wire to each portion of the circumference of the cup in the direction of a radius. The reaction between the radiating and the circular currents, causes the mercury conveying the former to revolve around the point of the wire, that being the centre from which the currents radiate. This rotation is rapid, and produces sufficient centrifugal force to make the mercury hollow in the centre. As the end of the wire is thus uncovered, the current is interrupted and the revolution ceases. When the mercury returns to its level, the current is renewed, to be again suspended as before The rotation is rendered more visible by putting a drop or two of muriatic acid on the mercury.

379. The circular currents in the helix may be regarded as made up of short rectilinear currents, each of which flows at right angles to the radius of the helix. Confining our attention to the action of a single one of these currents, it will be seen that two only of those passing through the mercury are parallel with it. Of these, one is moving in the same direction, and is attracted; the other, on the opposite side of the inserted wire, moves in the reverse direction, and is repelled. These two forces conspire in giving a motion of rotation to those portions of the mercury which are acting as conductors. The two radiating currents which are precisely at right angles to the rectilinear one, and also all the oblique currents, are attracted and repelled by the rectilinear current, in such a manner that they tend to place themselves parallel with it. As the influence of every portion of the current in the helix urges each of the radiating currents to move in one and the same direction, a rapid revolution of the mercury conveying them is the consequence. The rotation of the mercury mentioned in § 285, is produced in a similar manner with that obtained in this instrument.

380. In connection with the subject of the attractions and repulsions of currents, it will be proper to give a slight sketch of the Electro-Dynamic Theory of Ampère. This theory ascribes the actions of currents upon each other, of magnets upon each other, and those exerted between currents and magnets, to one and the same cause, the mutual attractions and repulsions of electric currents. In a steel

magnet, the currents are regarded as circulating in a uniform direction, at right angles to the axis of the magnet, around the elementary magnetic particles (§ 249), forming closed circuits. The resultant action of such a system of currents would be the same as that of a series of currents flowing round the circumference of the magnet at each portion of its length and at right angles to the axis. This assimilates the action of a steel magnet to that of the electro-magnet, with the current circulating in its helix, and to the action of the helix itself. How such elementary currents are excited and maintained within the substance of a good conductor of electricity, like steel, has not been ascertained. It is supposed that the currents are flowing at all times around the particles of every body susceptible of magnetism, but that they neutralize the influence of each other, being turned indiscriminately in every direction. By the process of charging, the currents are brought into a uniform direction, and magnetic power is developed.

381. Ampère regards the magnetism of the earth itself as due to electric currents circulating within it, from east to west, in planes parallel, or nearly so, to the magnetic equator. These currents are probably in great part thermo-electric, due to the action of solar heat on successive portions of the surface; but there are also many circumstances favorable to the formation of galvanic circuits in the interior.

382. The theory of Ampère affords a simple explanation of the *tangential action* of a current,

removing the apparent anomaly with regard to the direction in which the force is exerted. When a conducting wire is brought near a magnetic needle and parallel to it, the current in the wire is flowing at a right angle to those in the magnet. In this position, their mutual action is the same as that described in § 379. The currents in the needle tend to arrange themselves parallel to that in the wire; and to effect this, the needle, if free to move, turns itself at right angles to the wire. The continued revolution of a conducting wire around a magnet, and of the magnet around the wire, depends upon the same principle as the rotation of one current under the influence of another. By substituting a helix in place of the magnet, the revolutions would occur in the same manner, on transmitting the current.

383. When two helices, in which currents are flowing in the same direction, are placed end to end, they attract each other like the individual turns of the coil represented in Fig. 160. If one of them is turned round so as to present its other extremity to the first, repulsion is exhibited, in consequence of the opposite directions of the currents. In two steel magnets, presented end to end, the currents, according to Ampère's theory, act in the same manner as those in the helices, and the poles attract or repel one another on the same principle.

384. We now proceed to consider the inductive action of currents, taking first in order those phenomena which are referable to the induction of a current on itself. When the poles of a small gal-

vanic battery, consisting of a single pair of plates, such as Smee's, or the sulphate of copper battery, are connected by a copper wire of a few inches in length, no spark is perceived when the connection is either formed or broken, or at most a very faint spark at the moment of opening the circuit; but if a wire forty or fifty feet long is employed, though no spark is seen when contact is made, a bright one appears whenever the connection is broken by lifting one end of the wire out of the cup in which it rests. By coiling the wire into a helix, the spark becomes more vivid.

385. The most advantageous length for producing the spark depends upon the diameter of the wire, and also upon the quantity and intensity of the battery. With a wire one sixteenth of an inch in diameter, and a single pair of Smee's, or the small sulphate of copper battery, a length of 50 or 100 feet will probably give the best result. An increased effect is obtained by employing a longer wire, of greater diameter. If a battery of higher intensity is used, such as a single pair of Grove's, the length of the wire may be still further increased. The greater is the quantity of the battery, the larger or shorter must be the wire, in order to transmit the whole current, and obtain the brightest spark. This peculiar action of a long conductor, either extended or coiled into a helix, in increasing the intensity of the current from a single galvanic pair, at the moment when it ceases to flow, was discovered by Professor Henry, in 1831.

386. With a wire two or three hundred feet long, a slight shock may be felt at the moment of opening the circuit, if its ends, near their connections with the poles, are grasped with moistened hands; with a shorter wire, shocks may be obtained through the tongue; their intensity increases until a length of five or six hundred feet is attained. A single pair of plates gives no shock directly; the peculiar and continuous sensation excited in the tongue when the current from a single pair is made to pass through it not being called a shock. With a battery of small size, or consisting of a number of pairs, greater lengths may be used with advantage for the shock, as well as for the spark. The maximum effects of a small battery are, as might be expected, much inferior to those of a large one. If the requisite lengths of wire are exceeded, the effects are lessened.

387. The brilliancy of the spark is greatly increased by employing a ribbon of sheet copper coiled into a flat spiral, instead of a wire. A description and figure of this instrument have been given in § 112. The spiral being connected with the battery, a brilliant spark will be seen, accompanied by a pretty loud snap, whenever contact is broken; and if two metallic handles are attached by wires to the cups of the coil, and held in the hands, a slight shock will be felt. If the battery is in feeble action, the shocks may be perceptible only when passed through the tongue. No shock can be obtained by interposing the body in the direct circuit with the coil, so that the battery current may traverse them in suc-

cession; as the electricity supplied by a single pai of plates is of too low intensity to be transmitted, to any considerable extent, by so poor a conductor as the human body. Professor Henry was the first to employ coils of metallic ribbon for obtaining sparks and shocks from a single pair of plates.

388. Clock-work Electrotome.—For the purpose of interrupting the circuit rapidly, in either a long wire or a spiral, the instrument represented in Fig. 165 is very convenient. In the cut, it is shown

Fig. 165.

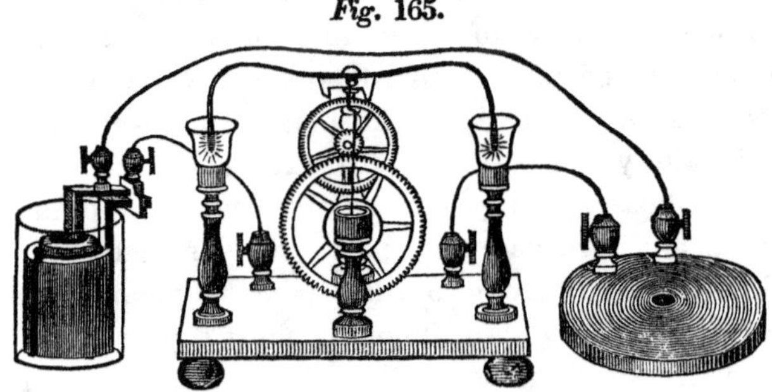

in connection with a ribbon spiral and a single pair of Grove's battery. It consists essentially of a bent copper wire, which, by means of clock-work set in motion by a spring, is made to vibrate rapidly, dipping its ends alternately into two glass cups, intended to contain mercury. The spring is wound up by turning a milled head. The glass cups are open at the bottom, to allow the mercury to come in contact with the brass pillars into which they are cemented These pillars are both connected with one of the screw-cups on the base board; the other screw-cup communicates with a brass cup for holding mercury on the top of a third pillar. Into this dips a vertical wire attached to the vibrating wire. Sufficient mercury must be put into the brass cup to keep the end of the vertical wire covered, and enough into the glass cups to allow one end of the vibrating wire to leave the mercury in its cup a little before the other end dips into its portion.

389. The clock-work electrotome may be advantageously used in connection with many of the instruments for affording sparks and shocks, which will be described subsequently. The current must be transmitted through the two instruments in succession, by connecting one of the screw-cups with one pole of the battery, and the other screw-cup with one of those attached to the spiral or other piece of apparatus, the remaining cup of which is to communicate with the other pole of the battery. It is better to break the circuit mechanically in this way, when it is desirable to obtain the best possible sparks and shocks, rather than by means of any interrupting apparatus worked by the battery itself, as a considerable part of the power of the current is then expended in giving motion to the interruptor.

390. On making connection with a flat spiral, in the manner shown in Fig. 165, and turning the milled head to put the vibrating wire in motion, a brilliant spark will be seen at each rupture of contact, accompanied by a loud snap, and causing combustion of the mercury at the point where the spark occurs. With a battery consisting of a few pairs of plates of

large size, the size of the spark will be greatly increased, and the snap become as loud as the report of a Leyden jar. The shock will also be pretty strong, and may be increased by covering the mercury in the glass cups with a stratum of oil. A shock may be obtained, especially when oil is used, on closing the circuit as well as on opening it, though inferior to that given in the latter case; a faint spark is also sometimes seen when the wire dips into the mercury.

391. The requisite length and thickness of the copper ribbon to give a maximum result depend upon the size of the battery employed. With spirals of considerable length, even if the copper be pretty thick, two or three pairs of plates are better than one, as the metal opposes some resistance to the passage of a current of low intensity. A ribbon spiral of moderate length interposed in the circuit of a compound battery, consisting of a considerable number of small pairs, produces scarcely any peculiar effect; while a coil containing three or four thousand feet of fine insulated wire will give an intense shock, though not a very brilliant spark, under the same circumstances. The higher the intensity of the electricity and the smaller its quantity, the less is the size requisite in the metallic conductor, and the greater may be its length.

392. The sparks and shocks given by long wires and by spirals are due to *secondary currents* induced in the metallic conductor at the moments of opening and closing the circuit. Their intensity is higher than that of the current which produces them, and

which, in this connection, is called the *primary current.* The phenomena belong to the same class as those presented by the secondaries induced in another conductor placed in the vicinity of the one which is conveying the battery current.

393. These secondary currents may be separated from the primary by placing a second flat spiral of copper ribbon over the one through which the battery current is transmitted, as represented in Fig. 166. Two wires being connected with the cups

Fig. 166.

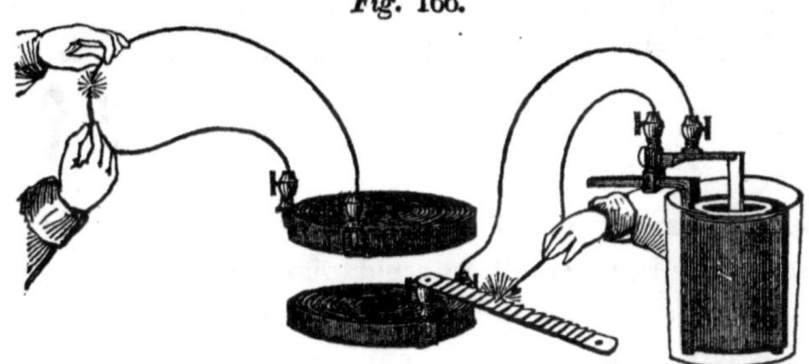

belonging to the upper spiral, rub their ends together while the circuit through the lower one is rapidly broken. Sparks will be seen, and slight shocks may be felt through the fingers, or by placing the wires in the mouth. When the ends of the wires are joined, the sparks and snaps given by the spiral connected with the battery are considerably diminished, and no shocks can be obtained from it.

394. Connect the cups of the upper coil with a delicate galvanometer, such as that represented in Fig. 39. Whenever the battery circuit is completed through the lower spiral, the magnetic needle is

deflected to a considerable extent, but immediately returns to the meridian, indicating the flow of a momentary current through the wire of the galvanometer. On opening the circuit, a similar transient deflection occurs in the opposite direction. There is no deflection while the battery current is flowing steadily. Care should be taken that the galvanometer is placed at such a distance from the lower spiral, that its needle may be unaffected by it. The spiral is able to deflect a delicate needle at a considerable distance, in the manner described in § 276, and if the galvanometer is found to be at all affected by making and breaking contact with the battery, when unconnected with the upper spiral, it must be farther removed.

395. A sewing needle will be magnetized if placed within a helix of small internal diameter connected with the upper spiral. The polarity produced by the current which attends the completion of the circuit, is the reverse of that communicated by the one attending its interruption. If both currents are allowed to act on the needle, it acquires little or no magnetism, as they flow in opposite directions, and neutralize each other's influence. The helix should consist of a single length of wire, wound so as to form eight or ten layers of coils, to enable it to be used with currents of considerable intensity. Its power will be greater if its internal diameter is very small.

396. As was stated in § 392, the momentary waves of electricity excited by electro-dynamic induction in

a conductor conveying a current, or in a neighboring one, are termed *secondary currents.* The wave which accompanies the closing of the circuit is termed the *initial secondary*, and flows in the opposite direction to that of the current which induces it. The other, which follows the opening of the circuit, is called the *terminal secondary*, and flows in the same direction as the inducing current. These currents were discovered by Professor Faraday, in 1831. They are, probably, occasioned by the disturbance of the natural electricity of the wire by the influence of the primary current.

397. In Fig. 167, a coil of fine insulated wire, W is represented placed over a flat spiral, A, one of

Fig. 167.

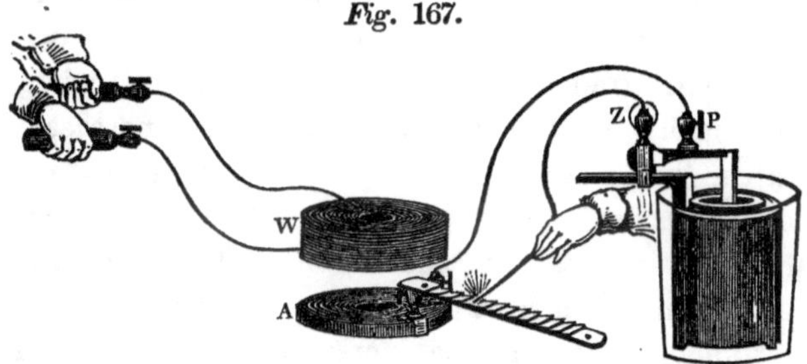

whose screw-cups is connected by a wire with the cup P, attached to the platinum plate of a single pair of Grove's battery. A wire from the cup Z, belonging to the zinc plate, is drawn over a steel rasp resting on the other cup of the flat spiral, for the purpose of breaking the circuit rapidly.

398. The ends of the wire coil, W, being fixed in the screw-cups of the metallic handles shown in the

cut, powerful shocks will be felt when these are grasped in the hands, and the wire connected with Z drawn over the rasp. In order to obtain the *initial* and *terminal* shocks separately, the current should be broken, not by means of the rasp, but by a cup containing mercury, into which one of the battery wires can be dipped at pleasure. The mercury should, of course, be connected by a wire with one of the cups attached to the flat spiral.

399. When a battery of a single pair of plates is employed, the initial secondary is much inferior in intensity to the terminal, and consequently gives a feebler shock. Professor Henry discovered that the intensity of the terminal current is very little increased by adding to the number of pairs; the slight increase which occurs is due to the greater quantity of electricity transmitted by the ribbon spiral, when the intensity of the battery current is increased. With the initial secondary it is different; every additional pair is found to raise its intensity, so that with about ten pairs it equals, in this respect, the terminal, and with a larger number excels it. The initial shock may also be increased, though not in any great degree, by employing a shorter flat spiral, as, for instance, one fifteen or twenty feet in length, with a single pair of plates. In *quantity*, as indicated by the galvanometer, the two secondaries are equal; those of the wire coil being inferior in this respect to the currents afforded by a ribbon coil.

400. The coil represented at W contains three thousand feet of copper wire, about one-fiftieth of an

inch in diameter, wound with thread; the layers are firmly cemented together by shellac, careful insulation being requisite, in consequence of the length of the wire and the high intensity of the current obtained. Where a small battery is used, this length of wire is unnecessary, as the shock given by it is scarcely greater than that from a coil of one thousand feet. With a larger battery, the longer one will be much superior. A sewing needle may be magnetized by the currents from a long wire coil, as well as by those from a flat spiral. If the wire is fine and very long, this effect will be diminished.

401. Instead of flat coils, long helices of insulated wire may be employed for obtaining the secondary currents, though with less effect when not aided by magneto-electric induction. Several of the magneto-electric instruments, which will be described under the next head, may be used for this purpose, the iron bar or bundle of wires being withdrawn from the helices. A description of one (the double helix and electrotome) may properly be introduced in this connection.

402. Double Helix and Electrotome. — In this instrument, represented in Fig. 168, the double helix, *a a*, is confined to the base board by three brass bands. The inner helix is composed of several strands of large insulated copper wire. The similar ends of these strands at one extremity of the helix are connected with the screw-cup *c*. Their other ends are soldered to the middle brass band, which is surmounted by a brass cup, *e*, for holding mercury.

Into this cup descends a copper wire attached to the wire *w w*, which, by means of clock-work set in

Fig. 168.

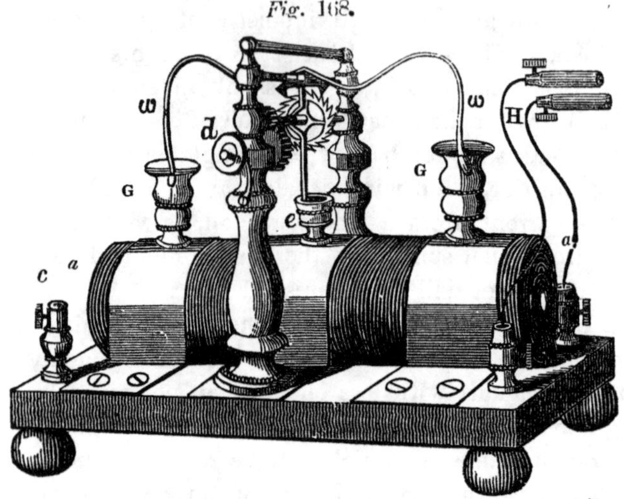

motion by a concealed spring, is made to dip its ends alternately into the glass cups, G, G, which are to contain mercury. The cups being open at the bottom, the mercury is brought in connection with the outer brass bands, upon which they are fixed. Both these bands are connected with a screw-cup *c′*, corresponding to *c*, but not seen in the cut. A second helix, consisting of about two thousand feet of fine insulated wire, encloses the one just described, but is insulated from it; its ends are soldered to the screw-cups to which the handles, seen at H, are attached.

403. The cups *c* and *c′* being connected with the galvanic battery, the current will pass through the nner helix whenever either end of the wire *w w* dips into the mercury, which should stand at such height in the cups that both extremities of the wire hall not be immersed at the same time. By turning the milled head, *d*, the spring is wound up, and the wire is made to vibrate rapidly. When either end leaves the mercury, the flow of the current is interrupted, and a bright spark is seen in the cup. If the handles are grasped with moistened hands, strong shocks will be felt whenever the circuit is broken. Introduce into the helix a brass tube, and the spark becomes small and the shock feeble; if the tube is sawn open in the direction of its length, it no longer produces these effects.

404. When an iron bar or a bundle of soft iron wires is introduced into the helix, the brass tube being withdrawn, the brilliancy of the sparks and the intensity of the shocks are greatly increased, the instrument being, under these circumstances, one of the most powerful belonging to the department of magneto-electricity.

405. We have seen that a battery current of considerable quantity and low intensity can induce either a quantity or an intensity current. By substituting for the ribbon spiral, through which the battery current is transmitted, a coil consisting of one thousand feet or more of fine insulated wire, and connected with a battery of a number of pairs, it will be found that an intensity current is able to induce secondaries of intensity in a wire coil, and of quantity in a ribbon coil.

406. The shocks obtained when the body is introduced into the circuit of a voltaic battery of a considerable number of pairs, without a coil, appear to be due to secondary currents induced in the battery itself. During the uninterrupted circulation of the galvanic current through the body, little or no effect is perceived; but at the moment of either opening or closing the circuit, a shock is experienced. When the series is very extensive, a dull pain is felt during the continuance of contact. The primary current has sufficient intensity to traverse the body, though not to give shocks, and doubtless induces *initial* and *terminal* secondaries when it commences and ceases to flow.

407. A flat spiral being in connection with the battery, let a fine wire coil be placed at a little distance above it; shocks may now be obtained from the wire, but their intensity diminishes in a rapid ratio as the distance between the coils is increased. With the arrangement represented in Fig. 169, shocks through the tongue are readily obtained when the wire coil is a foot or two above the other; and the distance may be still further increased by using a larger ribbon coil or a more powerful battery. This furnishes a convenient mode of regulating the intensity of the shock at pleasure; the same effect is produced when one coil lies upon the other, by sliding the wire coil from its central position more or less beyond the edge of the flat spiral. The shocks are in any case much increased by wetting the hands, especially with salt water. If, however, the

quantity of the secondary current is very small, the shock is lessened by improving the conducting power of the skin. This may happen when the secondary coil consists of very fine wire and the battery current is feeble.

408. The intensity of the shocks diminishes rapidly as the wire coil is raised from a horizontal position into an inclined one; and when it reaches a vertical position, its edge resting on the ribbon coil, they are no longer felt. Similar phenomena are presented when the flat spiral has a sufficiently large central opening to allow the wire coil to pass within it; no shocks being obtained when their axes are at right angles to each other. If the diameter of the wire coil is considerably less than that of the opening, and it is placed in a horizontal position within the spiral, the shocks are somewhat stronger when it is near the side than when in the centre.

409. The interposition of any good conductor of electricity between the fine wire coil and the one connected with the battery nearly neutralizes the shocks. The coils being arranged as represented in Fig. 169, interpose a slip of wood or a plate of glass between A and W, and the shock will be the same as if air only intervened. This will be the case with any non-conductor of electricity. Now, interpose a plate of metal, for instance,

Fig. 169.

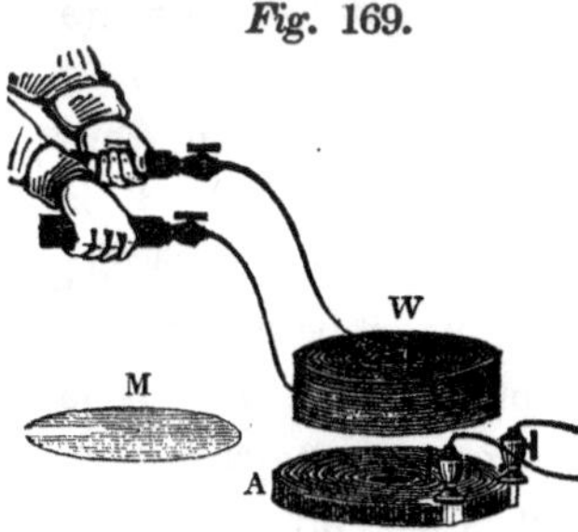

lead or zinc, one tenth of an inch thick, and as broad as the coils. The shock will be so much reduced as to be scarcely perceptible. If the interposed plate is broader than the coils, the reduction in the shock is still greater. The magnetizing power of the current is also lessened, in respect to hard steel; thus, a sewing needle, placed within a helix, will be but feebly charged. A certain thickness of metal is required to produce these effects, as several sheets of tin foil may be interposed without diminishing the shocks in any appreciable degree.

410. The interposition of a metallic plate does not prevent the occurrence of the secondary currents, but causes a great reduction in their intensity. That the quantity of the current is not affected, may be shown by connecting the ends of the upper coil, especially if it be a ribbon coil instead of a wire one, with a galvanometer. The deflections will be the same whether the plate is interposed or not, provided the distance between the two coils is not altered; but when the plate is of iron, the deflections are somewhat diminished.

411. In Fig. 169, M represents a metal plate, from which a slip is cut out in the direction of a radius, the cut extending to the centre. When such a plate is interposed between the coils, the shocks are not at all lessened. Instead of a metallic plate, interpose a flat spiral between the battery coil and the wire one. No diminution of the shocks will be perceived. Now connect the cups of the interposed coil by a wire, and the intensity of the shocks will be even more

reduced than by a plate of metal. Whenever the shocks are diminished, the brilliancy of the sparks given by the battery spiral are also lessened to some extent.

412. Secondary currents may also be obtained, without breaking the primary circuit, by altering the quantity of the battery current or the distance between the coils. A flat spiral being connected with a single pair of plates, place a second spiral of the same kind upon it, with its cups in connection with a galvanometer. While the current is flowing steadily through the lower spiral, no secondary is excited, and the needle of the galvanometer is unaffected. Now, lift one of the plates of the battery partly out of the liquid. The moment the plate begins to be raised, the needle moves in the same direction as if the circuit were broken; the deflection, however, is not momentary, as in that case, but continues during the movement of the plate. Then, without taking the plate out of the solution, which would break the circuit, depress it again. The galvanometer will now indicate a current in the opposite direction to the former one.

413. Similar currents are produced by raising the upper coil from the lower one, through which the galvanic current is steadily flowing. As the coil recedes, a secondary flows through it in the same direction as that of the battery current in the other spiral; as it again approaches, a current in the reverse direction is induced. Instead of raising the upper spiral, it may be moved laterally from its

central position on the lower one with the same result.

414. These currents produce a greater effect upon the galvanometer than those excited by closing and opening the circuit, as they are not momentary, but last as long as the motion continues. The more rapid the movement of the battery plate, or of the spiral, the more powerful are the secondary currents, as they depend upon the suddenness of the change in the quantity of the primary current in the one case, and in the distance between the coils in the other. They are, however, of low intensity, and are unable to afford shocks. The interposition of metallic plates or coils produces no effect upon them.

415. The neutralizing action described in § 409 and § 411, is due to a secondary current excited in the interposed metallic plate or spiral, which itself induces a *tertiary* current in the wire coil, flowing in an opposite direction to the secondary induced in it by the battery current, and therefore retarding its development. A tertiary current is also induced in the battery coil, which occasions the reduction in the spark and shock noticed in § 393 and § 411. When the interposed plate of metal is divided to its centre, no secondary is induced in it, and it exerts no neutralizing action; the same is the case with the ribbon spiral in § 411, when its cups are disconnected. Similar phenomena are produced by the introduction of a metallic tube into a wire helix, as described in § 403.

416. This *tertiary* current can be separated from

the secondary, and obtained by itself, in the following manner. A ribbon coil B (Fig. 170) being laid upon

Fig. 170.

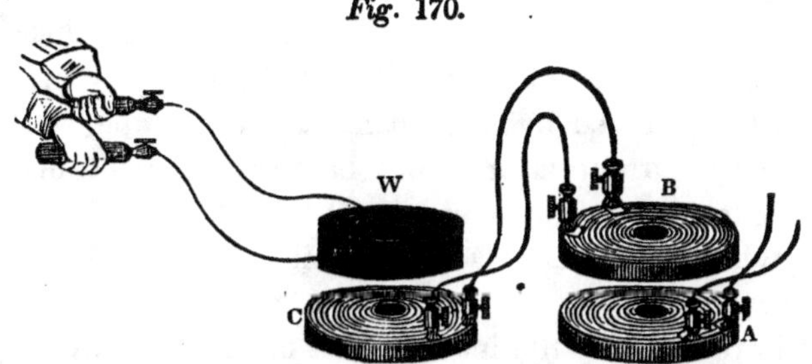

the coil A, through which the battery current is transmitted, connect its cups with those of a third spiral, C, of the same kind, removed to a little distance, so as to be beyond the influence of the current in A. The secondary current induced in B will now flow through C, and if a fine wire coil, W, is laid on C, strong shocks may be obtained. If W is raised up, the shocks are still felt when it is at a considerable height above C.

417. By placing a fourth ribbon coil on C instead of the wire coil, a quantity current will be obtained, capable of affecting the galvanometer slightly, and of magnetizing a sewing needle placed in a helix of small internal diameter. Let two fine wire coils be substituted for the flat spirals B and C. A secondary intensity current will now be obtained, which will induce a tertiary of intensity in a third wire coil laid on the second, enabling it to afford strong shocks, or a tertiary of quantity in a ribbon coil.

418. If the second spiral, B, is alone replaced by a wire coil, little or no shock can be obtained from W,

the quantity of the secondary current furnished by the wire coil not being sufficient for the production of a powerful tertiary, unless it is passed through a conductor of many convolutions. So, on the other hand, if a fine wire coil is substituted for C only, no tertiary is induced by it, or at most a feeble one, the secondary current from **B** not having sufficient intensity to enable it to overcome the resistance of the long wire. The tertiary current, like the secondary, can be induced at a distance, and has its intensity greatly reduced by the interposition of metal between the flat spiral C and the wire coil.

419. The tertiary currents may be conveniently obtained by causing the secondary from a ribbon spiral to flow through the inner helix of the instrument represented in Fig. 168, or of almost any of the magneto-electric instruments, which will be described under the next head. Thus, if the wires attached to B, in Fig. 170, are fixed in the cups *c* and *c'* of the double helix and electrotome, strong shocks may be obtained from the tertiary current induced in the fine wire helix. The circuit through the inner coil should not be broken by the electrotome, as the only interruption wanted is that in the battery current. The shocks are increased by placing a bundle of iron wires within the helix, as the inductive action of the current is then assisted by that of the electro-magnet.

420. Tertiary currents, like secondaries, are induced both when the primary circuit is opened and when it is closed. The *initial* and *terminal tertiaries* both flow in the opposite directions to the corresponding secondaries. In fact, each secondary must

produce two tertiaries, one when it commences, and another when it ceases to flow; but in consequence of the exceedingly short duration of the secondary itself, they cannot be separated as the initial and terminal secondaries can; and the current which is obtained is only the difference between the two. This accounts for the slight effect it produces upon the galvanometer, while capable of affording strong shocks. The two parts may differ very much in intensity, but, being equal in quantity, would not affect the galvanometer, did they occur precisely at the same instant. The needle, however, is first deflected by the momentary wave induced by the commencement of the secondary, and, as soon as it has moved a degree or two, is arrested by the wave due to its cessation, and carried in the opposite direction.

421. The effects of the interpositions described in § 409 and § 411 may now be more fully explained. The secondary induced in the interposed conductor, on opening the primary circuit, itself induces a tertiary in the wire coil at the instant of its commencement, which flows against the secondary induced in it by the battery current. When the secondary in the interposed body ceases, another tertiary is excited in the wire coil flowing in the same direction as the secondary. The total amount of the current is not altered, since the same quantity is added at its ending as was subtracted at its beginning; but its intensity is greatly reduced, probably in consequence of the diminished rapidity of its development.

422. Currents of higher Orders.—It has been shown that a secondary current, though only momentary in its duration, can induce a tertiary of considerable energy. It might therefore be expected that the tertiary would produce a current of the fourth order; this another, and so on; and such is found to be the case. It is only necessary to remove the tertiary out of the influence of the secondary, in the same manner as the secondary is removed from that of the primary (see § 416), in order to obtain a current of the *fourth order.* The currents of the third, fourth, and fifth orders were first obtained by Professor Henry, and two other orders have been since added. These currents progressively diminish in energy, but the phenomena presented by them are similar to those of the tertiary. With a larger number of coils and a powerful battery, the series might doubtless be extended much farther.

423. In the following table, the directions of the currents produced both at the beginning and ending of the battery current are given; those which flow in the same direction as the primary being indicated by the sign +, and those in the opposite direction by the sign —.

	At the beginning.	*At the ending.*
Primary current,	+	+
Secondary current,	—	+
Tertiary current,	+	—
Current of the fourth order,	—	+
Current of the fifth order,	+	—
Current of the sixth order,	—	+
Current of the seventh order,	+	—

If the induction at the ending of the battery current is regarded as opposite to that at the beginning, the second column may commence with minus instead of plus, and the second series will then alternate like the first.

424. Induced currents of the different orders may be obtained from frictional electricity, though, in consequence of its high intensity, the conductors require better insulation than is necessary when they are used with the galvanic battery. The flat spirals and wire coils may, however, be employed, if their layers are carefully insulated by means of shellac, or if covered with silk instead of cotton.

425. Let a fine wire coil be placed over a ribbon spiral, with a plate of glass interposed; a secondary shock may now be obtained from the wire when the charge of a Leyden jar is passed through the spiral. A still better mode is to employ a second wire coil, instead of the flat spiral; if the ends of one of the coils are held in the hands, a strong shock will be felt at the moment of discharging the jar through the other. The secondary current flows in the same direction as the one which induces it; as may be shown by passing it through a helix consisting of several layers of coils, when it will magnetize a sewing needle placed within it.

II. BY THE INFLUENCE OF A MAGNET.

426. The name of *Magneto-Electricity* is given to that branch of science which treats of the development of electricity by the influence of magnetism. Electric currents are excited in a conductor of electricity by magnetic changes taking place in its vicinity. Thus the movement of a magnet near a metallic wire, or near an iron bar enclosed in a wire coil, occasions currents in the wire. They are also produced at the moment of commencing and suspending the flow of a galvanic current through the coil surrounding an electro-magnet. In this case, an induced current may be obtained either from the wire conveying the primary current, or from a second wire also surrounding the iron; but the current excited by the influence of the magnetized bar is obtained in connection with that which is the result of electro-dynamic induction, and cannot be separated from it, at least with the usual arrangement of the wires. The discovery of the induction of electricity, both by steel magnets and by electro-magnets, was made by Professor Faraday, in 1831, during the same course of experiments which led to the discovery of electro-dynamic induction.

427. Let the heliacal ring (§ 271) be connected with a gold leaf galvanoscope, as represented in Fig. 171. By passing the ring over one of the poles of a steel U-magnet, the gold leaf will be sensibly

deflected during the continuance of the motion. On withdrawing the ring from the pole, a deflection in the opposite direction will occur, and in the same direction as that obtained by passing it over the other pole. The heliacal ring may be connected with a delicate galvanometer, such as is described in § 162, with the same result.

Fig. 171.

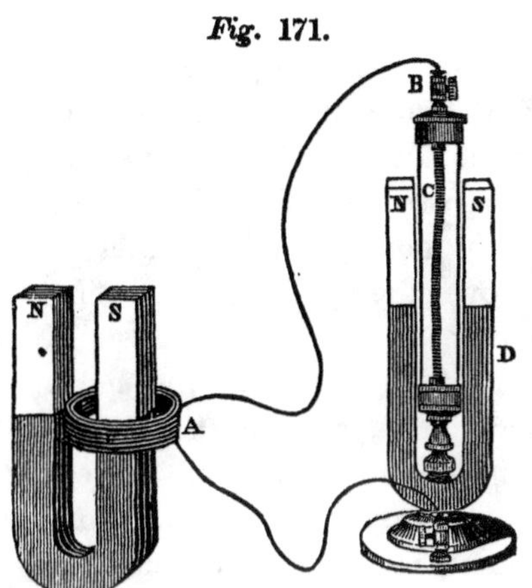

428. If the helix represented in Fig. 107 is connected with the galvanometer, and a bar magnet or one pole of a U-magnet is introduced into it, the needle of the galvanometer is deflected while the magnet is passing in, but returns to its former position as soon as the magnet is at rest within the coil. On drawing the magnet out, the needle is deflected in the opposite direction. By moving the magnet in and out so as to keep time with the oscillations of the needle, they are greatly increased. Reversing the direction of the magnet, so as to make it enter by the contrary pole, reverses the indications of the galvanometer. If the bar magnet is carried through the helix, and brought out at the opposite end to that by which it entered, the effect is the same as if it had been draw

out at the same end. No current is excited while the magnet and helix are both at rest.

429. Connect the cups of a flat spiral (Fig. 112) with the galvanometer, and pass a U-magnet over it, towards the centre, with one of its poles above and the other below. The needle will be deflected in opposite directions as it passes on and off. A less effect will be produced by moving a bar magnet in the direction of a radius over the spiral, or by passing it into the central opening.

430. Place a bar of soft iron within the helix (Fig. 107), and connect its cups with those of a galvanometer. Then bring either pole of a steel magnet in contact with one extremity of the iron. The bar instantly becomes magnetic, and the galvanometer needle is deflected by the current produced by this means in the helix. It, however, immediately returns to its former position, the settled magnetic condition of the bar having no power to excite a current. On withdrawing the magnetic pole, the bar loses its magnetism, and the needle is deflected in the opposite direction.

431. By alternately bringing the magnet in contact with the iron, and withdrawing it in such a manner as to keep time with the vibrations of the needle, they may be greatly increased. If the opposite poles of two bar magnets are brought in contact with the ends of the iron bar, and especially if their other poles are made to touch, so that the two magnets form the letter V, the inductive influence is much greater. By opening and shutting the magnets, as

if joined by a hinge at the vertex, the magnetism of the bar within the helix may be communicated and removed at pleasure.

432. When an armature or any piece of soft iron is brought in contact with one or both of the poles of a magnet, it becomes itself magnetic by induction, and by its reaction adds to the power of the magnet. On the contrary, when taken away, it diminishes the power of the magnet. This alteration in its magnetic state induces a current of electricity in a coil surrounding it, as may be shown by passing a wire coil, whose ends are connected with a galvanometer, over one of the poles of a U-magnet, as represented in Fig. 171, keeping the magnet and coil stationary. The needle will now be deflected in one direction when an armature is applied to the poles, and in the opposite direction when it is removed.

433. The most powerful effects are obtained by causing a bar of soft iron, enclosed in a helix, to revolve, by mechanical means, near the poles of a steel magnet. The principle may be illustrated by connecting the cups, A and B, of the revolving electro-magnet (Fig. 146) with a delicate galvanometer, as represented in Fig. 172. On causing the iron bar to rotate rapidly, by drawing the hand over the axis, the needle is powerfully deflected. As the iron approaches the poles of the steel magnet, it becomes magnetic; as it recedes from them, its magnetism disappears. While one end of the bar is leaving the north pole, or approaching the south pole, the electric current flows in a constant direction, the loss of south polarity by the iron, and the production of north polarity in it, inducing currents in the same direc-

Fig. 172.

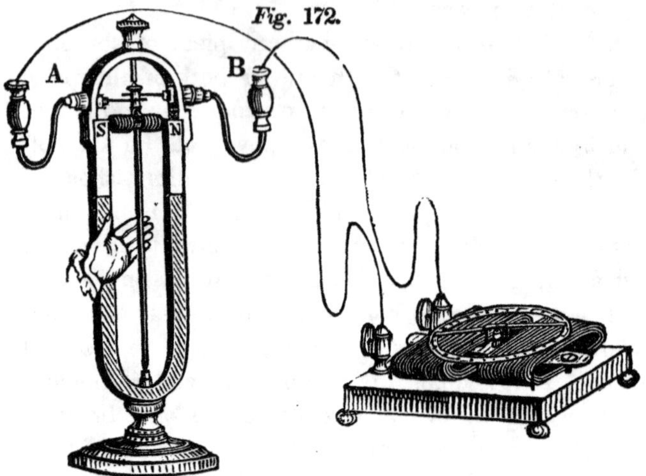

tion. During the other half of the revolution, the same end of the bar is approaching the north pole or receding from the south, and a reverse current is produced. The two currents are conveyed to the galvanometer in one direction by means of the pole-changer on the axis of the revolving bar. The pole-changer, in place of altering the direction of a current conveyed to the coil, as it does when the bar is made to revolve by a galvanic battery, here performs the duty of bringing two opposite currents from the coil into one course. The current traversing the wire of the galvanometer is, consequently, not reversed, except by changing the direction of the rotation.

434. If the cups A and B are connected with the cups *c* and *c′* of the double helix and electrotome, slight secondary shocks, which may sometimes be felt in the hands, will be obtained from the fine wire helix by rotating the iron bar, as in the cut. The hollow of the double helix should be filled with iron wires, and the vibrating wire be put in motion so as to break the circuit rapidly.

435. With a sufficiently delicate galvanometer, any of the electro-magnetic instruments in which motion is produced by the mutual action between a galvanic current and a steel magnet, may be made to afford a magneto-electric current by producing the motion mechanically. In all cases, the current excited flows in the opposite direction to the galvanic current which would be required to produce the same motion.

436. When a galvanometer is used in these experiments, it must be placed at such a distance from the instrument where the movements and magnetic changes are made, that the needle shall not be deflected by any influence but that which reaches it through the connecting wires. With the gold leaf galvanoscope this precaution is unnecessary.

437. Magneto-Electric Machine.—In the instrument represented in Fig. 173, an armature, bent twice at right angles, is made to revolve rapidly, in front of the poles of a compound steel magnet of the U form. The U-magnet, whose north pole is seen at N, is fixed in a horizontal position, with its poles as near the ends of the armature as will allow the

latter to rotate without coming in contact with them. The armature is mounted on an axis, extending from

Fig. 173.

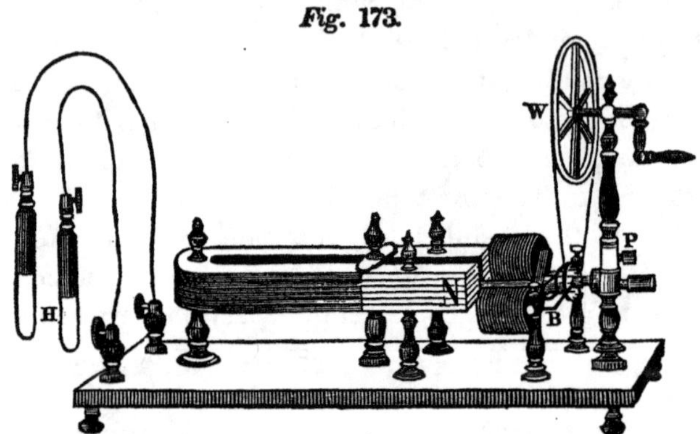

the pillar P to a small pillar between the poles of the magnet. Each of its legs is enclosed in a helix of fine insulated wire. The upper part of the pillar P slides over the lower part, and can be fastened in any position by a binding screw. In this way the band connecting the two wheels may be tightened at pleasure by increasing the distance between them. This arrangement also renders the instrument more portable than would otherwise be the case.

438. By means of the multiplying wheel W, which is connected by the band with a small wheel on the axis, the armature is made to revolve rapidly, so that the magnetism induced in it by the steel magnet is alternately destroyed and renewed in a reverse direction to the previous one. When the legs of the armature are approaching the magnet, the one opposite the north pole acquires south polarity, and

the other north polarity. (See § 228.) The magnetic power is the greatest while the armature is passing in front of the poles. It gradually diminishes as the armature leaves this position, and nearly disappears when it stands at right angles with the magnet. As each leg of the armature approaches the other pole of the U-magnet, by the continuance of the motion, magnetism is again induced in it, but in the reverse direction to the previous one.

439. These changes in the magnetic state of the armature excite electric currents in the surrounding helices, powerful in proportion to the rapidity with which the magnetic changes are produced. The helices are so connected as to form a continuous coil. The ends of this coil are soldered to the segments of a pole-changer (Fig. 70), secured on the axis. Two silver springs press upon these segments, and convey the electricity to the two screw-cups. Although the currents in the helices are reversed twice in each revolution, they are turned into one direction by means of the pole-changer. From the manner in which they are obtained, they necessarily vary more or less in power in the different parts of the revolution, according to the position of the armature.

440. This *primary* magneto-electric current has too low an intensity to afford strong shocks. But *secondary* currents may be obtained by interrupting the primary circuit, as with the galvanic current. (See § 392.) These have a much higher intensity, and give powerful shocks. One of the springs press

ing on the pole-changer is fixed by a binding screw in B, and the other in a similar pillar opposite to B. They are bent at right angles, and when both the horizontal and vertical portions are made to touch the segments of the pole-changer, the circuit is broken as the armature revolves. The horizontal parts simply bring the two primary currents into one course. For showing the sparks, another wire-spring, fixed in the pillar, B, is made to play upon steel pins set in the small wheel on the shaft.

441. Shocks are obtained at every part of the revolution; but, with a moderate speed, they are most powerful when the legs of the armature are near the magnet. If the motion is very rapid, this difference is less appreciable, and with a powerful machine, the torrent of shocks which then results becomes insupportable. The muscles of the hands which grasp the handles are involuntarily contracted, so that it is impossible to loosen the hold. The shocks are, however, instantly suspended by bringing the metallic handles into contact. A spark is seen when the wire fixed in the pillar B leaves each of the pins on the wheel.

442. In Fig. 173, the metallic handles are represented in connection with the screw-cups, for the purpose of giving shocks. When these are held in the hands, the arm connected with the negative cup will be found most affected by the shocks. This is a physiological phenomenon, the current producing a greater effect upon the arm in which it flows downwards, in the direction of the ramification of the

nerves, than upon the one in which it ascends. The initial secondary is too feeble to afford shocks, so that only the terminal secondary need be taken into account. The intensity of the terminal shock is, however, constantly varying, according to the position of the armature in respect to the magnet, and the difference in the effect upon the two arms is not so distinctly marked as with some of the instruments which will be described hereafter.

443. The shocks may be regulated to some extent, by varying the speed with which the armature is made to revolve. They are considerably lessened by partially neutralizing the power of the steel magnet, by placing an armature across it near the poles; also, by passing the current through a piece of wet cotton wicking, a few inches long, one end of which is connected by a wire with one of the screw-cups on the base board, from which the handle is removed. By grasping in one hand the handle connected with the other screw-cup, and in the remaining hand the disconnected handle, and touching a short wire attached to this to the wet cotton, the shock is diminished in proportion to the length of the imperfect conductor which the current is obliged to traverse. The handles represented in Fig. 173 are of German silver. Between the metallic part of each handle and the screw-cup attached to it, is interposed a cylinder of wood. No shock is felt when one or both are held by the wooden portions.

444. Slight shocks may be obtained from the primary current, by grasping the metallic handles

connected with the screw-cups. The wire at B, which plays upon the pins, must be removed, so that the circuit shall not be broken. It is also essential that neither of the springs pressing on the pole-changer should leave the segment which it touches before it comes in contact with the opposite segment. If this is neglected, the circuit will be interrupted at the pole-changer, and strong shocks obtained. When the screw-cups are connected with those belonging to the inner coil of the double helix and electrotome (§ 402), and the central opening of that instrument is filled with iron wires, secondary shocks of considerable strength will be obtained from the exterior helix whenever the armature is made to revolve. The vibrating wire should be put in motion to break the primary circuit. Bright sparks are at the same time seen in the mercury cups. The magneto-electric sparks are conveniently shown by passing the primary current of the machine through the clock-work electrotome (Fig. 165), the vibrating wire of that instrument being set in motion.

445. When the primary magneto-electric current is made to pass through water in a constant direction, the water is resolved into its elements, and the gases hydrogen and oxygen are given off separately, by the two wires which convey the current. Two platinum wires being connected with the screw-cups, and their ends immersed in water, a slender stream of gas will be seen to escape from each wire when the armature is made to revolve. The decomposing cell, represented in Fig. 28, is well adapted to this

experiment. A little sulphuric acid may be added to the water, but care should be taken not to make the conducting power too great, or the amount of gas evolved will be lessened.

446. The experiments described under the head of Galvano-Decomposition (§ 66 to § 70) may be performed with the magneto-electric machine, though the effects are on a smaller scale. The primary current is preferable to the secondary for this purpose. Strips of platinum foil, which are generally superior to wires in decomposing by a compound galvanic battery, do not answer so well with the magneto-electric current, especially when the wire coiled upon the armature is fine.

447. Let the decomposing tube (Fig. 31) be filled with a weak solution of iodide of potassium, without any coloring liquid. By causing the armature to revolve, iodine will be abundantly liberated round the positive wire; this, being slightly soluble, gives a brown color to the liquid, but most of it remains in suspension, forming a dense cloud. If a few drops of a weak solution of starch had been previously added, an intense blue color will be developed.

448. When a solution of sulphate of copper is employed, sulphuric acid and oxygen are set free in the positive cell, and metallic copper is precipitated upon the negative wire. If the current is powerful, it is deposited as a slightly adherent black powder; but if of moderate strength, a thin coating is formed, possessing the proper color and appearance of the

metal. In this case little or no hydrogen escapes from the coated wire, though oxygen is given off by the positive one. On reversing the current, the copper is gradually dissolved off from the coated wire, and a similar deposit occurs on the other. No oxygen escapes from the wire which is now positive, until its coating has nearly disappeared. When the experiment is concluded, the deposited copper may be removed from the platinum wires by a little diluted nitric acid. If two copper wires are immersed in the solution, as much copper will be dissolved off of one as is deposited upon the other. Sulphuric acid does not act upon copper in the cold, unless aided in this way by an electric current.

449. Let the tube contain a solution of cyanide of gold (§ 102), the conducting wires being of platinum. The negative wire will soon become covered with a coating of gold, which increases in thickness as the current is continued. Other metals, as, for instance, silver, copper, and brass, may be gilded by attaching them to the negative wire, and immersing them in the solution. The coating does not adhere firmly, unless the metallic surface on which it is to be deposited has been perfectly cleaned, as directed in § 103.

450. Many other metallic salts may be decomposed in the same manner, and the metals precipitated; but in most cases, the deposit is of a black color. Both wires may be in the same portion of liquid, no partition being required. The deposition of the metals from their solutions depends upon the same principles

which are concerned in the case of the galvanic current, as explained under the head of Electro-Metallurgy. The magneto-electric machine is sometimes employed in electro-gilding and silvering, in place of the galvanic battery. When the mechanical power requisite to maintain the motion of the armature can be conveniently obtained, it may be used with advantage; especially if the coils are constructed in the manner which will shortly be described, so as to furnish a current of considerable quantity.

451. The galvanometer is strongly affected by the primary magneto-electric current. Even a large and heavy needle, surrounded by a single wire, as in the instrument represented in Fig. 53, may readily be deflected. A sewing needle, or a piece of steel wire, placed in a magnetizing helix, will be fully charged. When the extremities of the wire surrounding a small electro-magnet, such as is represented in Fig. 126, are fixed in the screw-cups, it will sustain a weight of some ounces while the primary current is flowing. If the electro-magnet is covered with four or five layers of coils, the wire being in a single length, it will lift several pounds.

452. Improved Magneto-Electric Machine.— An improved form of the machine is represented in Fig. 174. Two straight armatures, surrounded by coils of insulated copper wire, revolve between two U-magnets, according to Dr. Page's plan; but much shorter armatures are used than in the machine contrived by him. The steel magnets are fixed, with the south pole of one above the north pole of the

other, at such a distance as just to allow the armatures to pass between them. These are mounted,

Fig. 174.

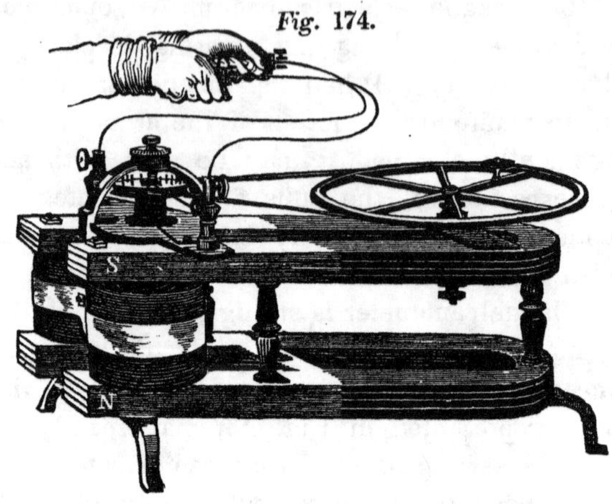

one on each side of a vertical shaft, in such a manner that both shall be passing between the opposite poles at the same time. They are made to revolve rapidly by a multiplying wheel, as in Fig. 173. A pole-changer on the shaft conveys the alternating currents in a constant direction to the screw-cups with which the metallic handles are shown in connection.

453. For giving shocks, the small wheel on the shaft is set with vertical pins, upon which plays an iron or steel wire connected with one of the screw-cups. Brilliant sparks are seen as the wire passes over the pins. These sparks are sometimes half an inch in length. The shocks are stronger than with the machine last described, and the decomposing power considerably greater.

454. Magneto-Electric Machine, with Vibrating Electrotome. — In this form, the circuit is not interrupted by the motion of the machine itself, for giving shocks; but the vibrating electrotome, which will be described immediately, is attached for that purpose, as represented in Fig. 175. The shocks

Fig. 175.

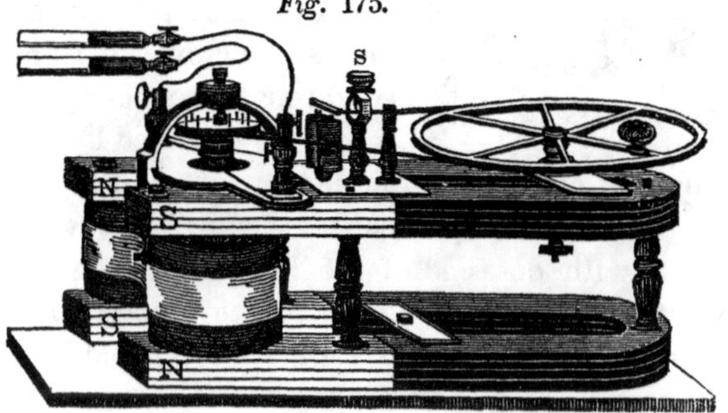

are of more uniform strength with this interruptor than with the mechanical one, as they occur only when the primary current charges the electro-magnet sufficiently to draw down its armature. For sparks, decompositions, &c., the primary current can be used without passing it through the coil of the electrotome. This is effected by turning the screw S, so as to raise the platinum point from the spring attached to the armature.

455. Vibrating Electrotome. — In Fig. 176, the vibrating electrotome is shown in a form adapted for use with either of the machines described in § 437 and § 452. There is an electro-magnet, of the U form, enclosed in a helix consisting of several layers of coils. Above this is a straight armature,

fixed to a spring, by which it is held up from the poles. The screw-cups at one extremity of the base board are to be connected with the cups of the machine, and those at the other end with the handles for shocks. The primary magneto-electric current is passed through the helix, and a secondary obtained by interrupting the circuit by the movement of the armature. One screw-cup at each extremity of the stand connects with one end of the helix, which consists of a rather long single wire, and the other two cups with the pillar to which the armature is fixed. The shock is thus obtained from the same wire which conveys the primary current.

Fig. 176.

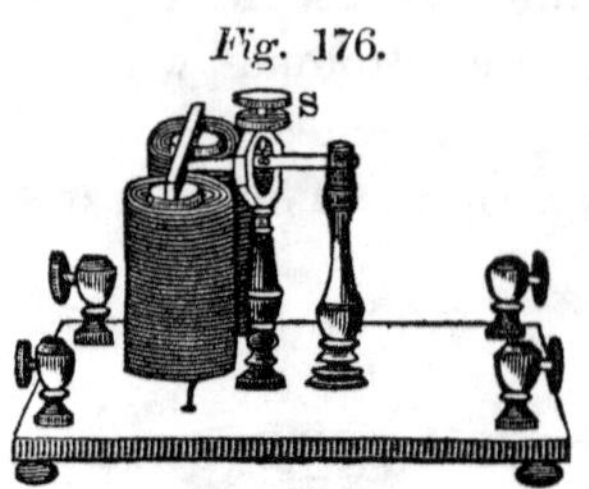

456. When the machine is put in motion, the electro-magnet is instantly charged, and draws down its armature. This motion depresses the spring, and separates a little platinum disc upon it from a point of the same metal attached to a set screw, S, above the spring. On the interruption of the circuit, the armature is raised by the force of the spring. A thin slip of brass brazed to the face of the armature prevents it from being retained by the poles. The shocks are varied to some extent, by turning the screw, S, which alters the distance between the armature and the poles.

457. The primary magneto-electric current resembles a galvanic current excited by a number of

small pairs. Its quantity and intensity are, however, both greatly influenced by the size and length of the wire enveloping the armature. A long and fine wire affords a current of small quantity and high intensity, and is most suitable for giving shocks. A short wire of large diameter gives a current of moderate intensity, but of considerable quantity, and is, therefore, best for producing sparks, decompositions, and magnetism.

458. Magneto-Electric Machine, for Quantity. — The machine represented in Fig. 177 is similar to the one described in § 452, except that the armatures are wound with short coils of coarse wire. By this arrangement, a quantity current is obtained,

Fig. 177.

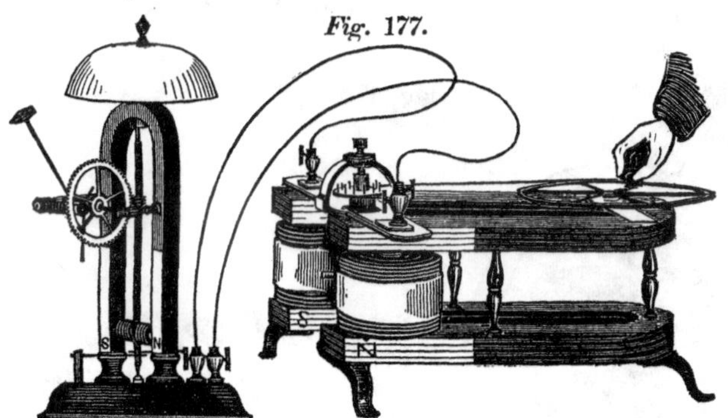

capable of producing motion in some of the electro-magnetic instruments, like the galvanic current. In the cut, the revolving bell engine (Fig. 147) is shown in connection with the machine. The primary current is employed, and those instruments are best suited to the purpose which do not require a powerful current to operate them. Both quantity and intensity armatures may be adapted to any of the machines, so that one may be substituted for the other at pleasure.

459 We now pass to a class of instruments in which electric currents are induced by electro-magnets whose magnetism is alternately acquired and lost. These instruments consist essentially of double helices containing bars or wires of soft iron. The magneto-electric current is thus obtained in conjunction with that excited by electro-dynamic induction, and the current formed by their union is called a *secondary*, though only in part such.

460. Platinum wire may be ignited by this secondary current in the manner represented in Fig.

Fig. 178.

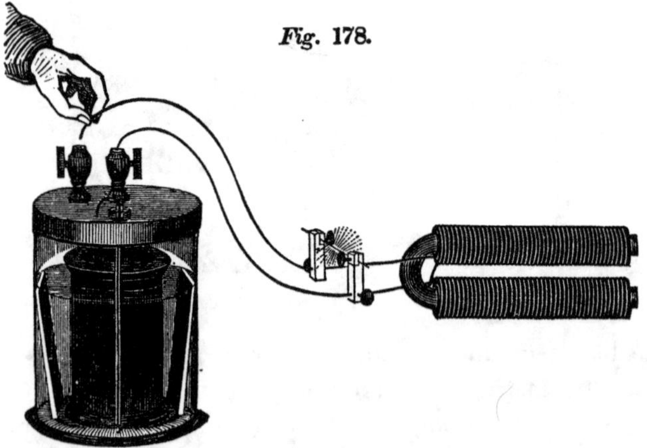

178. A short piece of fine platinum wire is secured by binding screws to the wires of a large electro-

magnet. The wires are uncovered at that point, and being connected by the platinum, a part of the battery current passes through it. It should be long enough not to be fully ignited by this current. When one of the wires is connected with one pole of a single pair of Grove's battery, and the remaining wire brought in contact with the other pole, the electro-magnet becomes charged. On breaking contact, its magnetism instantly disappears, and a current is induced in the surrounding coil which flows in conjunction with the secondary excited by the battery current itself. The combined currents have no circuit open to them except through the platinum wire, which is for the moment ignited by their passage.

461. Since the description of Smee's and Grove's batteries, in the earlier part of the volume, was written, some experiments have been made in relation to them, which are of sufficient interest to be mentioned here, though out of their proper place. The first trials were made to ascertain the relative power of a single pair of Smee's battery with the plates at different distances. The plates used were flat, and were placed opposite to each other, in a porcelain trough, containing sulphuric acid, diluted with 15 times its measure of water. In the whole series of experiments, the power was measured by the magnetism developed in a magnetometer (Fig. 131), through the coil of which the current was passed.

462. In the following table, the first column gives the distance of the plates in inches and fractions of

an inch. The second and third columns give the attractive force of the electro-magnet, or the magnetizing power of the current, in grains. The weights in the second column were obtained with a smooth platinum plate; those in the third, with one platinized in the usual manner.

TABLE I.

Distance.	*Grains.*	*Grains.*
10	25,000	30,000
5	30,000	34,000
2½	32,500	37,500
1¼	38,000	40,500
⅝	40,000	42,500

It will be seen that, as the plates are approximated, the power gradually increases, but not in any very great degree. The table shows the advantage of platinizing the negative plate.

463. With a battery of the form represented in Fig. 178, the increase of power by adding to the size of the plates was determined. The negative plate consists of one or more pieces of zinc, whose lower ends are placed in the mercury at the bottom of the battery. The weights in Table II. were obtained with the battery arranged as a Smee's, the porous cell being removed. In the first column is given the number of zinc plates, the whole being connected as a single plate. One piece of platinum was employed in the three first experiments, and in the fourth, two pieces, connected as one.

TABLE II.

Number of Zinc Plates.	*Grains.*
1	47,000
2	51,000
3	53,000
3	56,000

464. The battery was now converted into a Grove's, by using the porous cell, and its power with different exciting liquids examined. Equal measures of sulphuric acid and water were poured into the porous cell, and to this liquid was subsequently added pulverized nitre, in portions of one spoonful each. The results are given in the following table:—

TABLE III.

Fluid in Cell.	*Grains.*
Sulphuric acid and water,	50,000
With 1 portion of nitre,	65,000
" 2 portions " "	86,000
" 3 " " "	91,000

465. When the porous cell was filled with strong nitric acid, the magnetizing power was equal to 90,000 grs. The mixture of nitre and sulphuric acid, though inferior to strong nitric acid, makes a good exciting solution. Nitre, being a nitrate of potash, is decomposed when added to sulphuric acid, which forms sulphate of potash and liberates nitric

acid. Hence the active agent is here also nitric acid· but in using the mixture, the disagreeable fumes of nitrous acid produced by strong nitric acid alone are in a great degree avoided.

466. Two of Grove's batteries, of the form represented in Fig. 178, were connected with the magnetometer, at first united as a single pair and afterwards consecutively, strong nitric acid being used in the porous cell. The following table gives the magnetizing power obtained: —

TABLE IV.

Battery No. 1, ...	80,000 grs.
" No. 2, ...	82,000 "
Both, as one pair, ...	88,000 "
" consecutively,.	97,000 "

Owing to the magnetizing power being so great as nearly to reach the limit of saturation of the electro-magnet, the numbers in this table do not indicate so great an increase as really occurs; but they show that no great advantage is gained by connecting two pairs as a single pair. The large-sized Smee's or Grove's batteries are not much superior to the small ones. Ten pairs of Smee's battery, connected as one, are but little better than one pair.

467. SEPARABLE HELICES. — In this instrument, which is represented in Fig. 179, there are two helices entirely separate from each other. The inner one, composed of several strands of insulated coarse copper wire, is fixed in a vertical position on the base

board. One of its ends is connected with the screw-cup A, and the other with a steel rasp, B. The

Fig. 179.

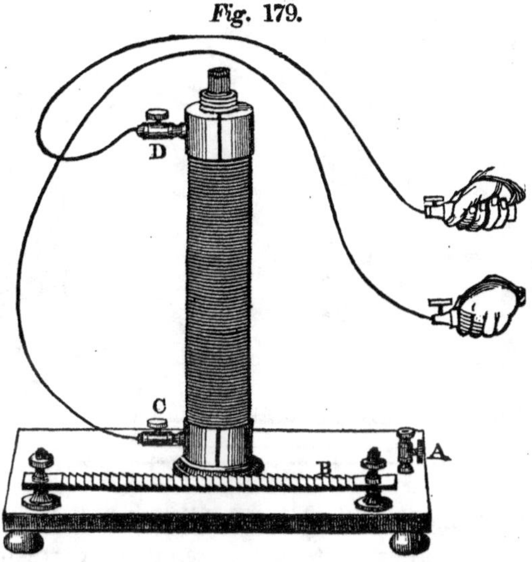

exterior helix is of fine insulated wire, and can be lifted off from the other, which it surrounds. Its ends are enclosed in two brass caps, to which the extremities of the wire are soldered. To these caps are attached the screw-cups C and D. A bundle of annealed iron wires, of which the ends are seen in the cut, can be removed from the inner helix when desired.

468. In Fig. 180, the different parts of the instrument are shown separately. The exterior helix, *a*, is removed from the inner coil, *b*, which is fixed to the base board. At *c* is seen a brass tube, within which

is the bundle of iron wires, *d*, intended to be introduced into the interior helix. For giving the strong-

Fig. 180.

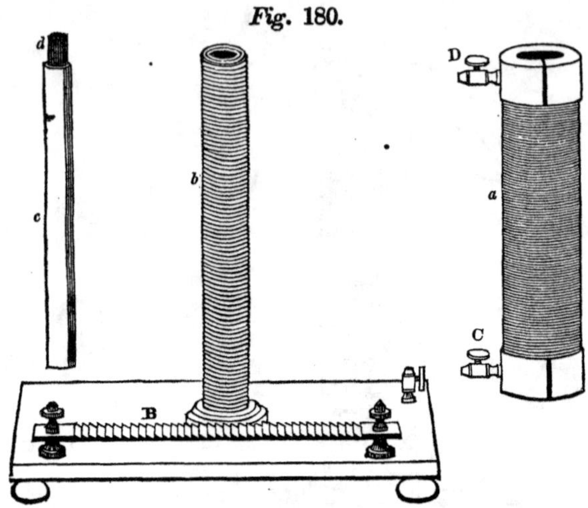

est shocks, the bundle should fill the hollow of the helix. The other parts are lettered in correspondence with the last figure.

469. The bundle of iron wires being withdrawn, let a wire connected with one pole of a galvanic battery be fixed in the cup A, and the other battery wire be drawn over the steel rasp. Bright sparks will be seen, and if metallic handles connected with C and D are grasped in the hands, as represented in Fig. 179, slight shocks will be felt on completing the circuit at the rasp, and stronger ones when it is broken, as with the instrument described in § 402, which is on the same principle.

470. If a rod of soft iron is introduced into the

helix, the spark is much increased, brilliant scintillations are produced, and the shock, when the circuit is broken, becomes powerful. The iron acquires and loses magnetism whenever the current begins or ceases to flow, and induces secondary currents in both of the coils which surround it. In the coarse wire coil, which conveys the battery current, this appears in the increased sparks and scintillations. In the fine wire coil it is felt in the strong shock which results.

471. When the bundle of iron wires is substituted for the soft iron rod, the spark and shock are much greater. If the rod or bundle of wires is introduced gradually into the helix, the spark and shock increase as it enters. The intensity of the shock may be varied at pleasure, by altering the number of iron wires, the addition of a single wire producing a manifest effect. If a glass tube is slipped over the iron wires in the helix, it does not interfere with their inductive action on the surrounding coils. But if a brass tube is passed over them, their influence is entirely suspended, so far as the shock and the spark are concerned. When the tube is slipped partly over them, their influence is partially suspended. This also is a means of regulating the shock without altering the battery current.

472. The neutralizing action of the tube is thus explained. The magnet induces in the tube, as well as in the two coils, a secondary electric current, which flows around it when the circuit is completed or broken. This secondary induces a tertiary cur-

rent in each of the coils, which flows at the first instant in an opposite direction to the secondary induced in the coil by the magnet, and therefore retards it. As the secondary current in the tube is, however, instantaneous, it induces another tertiary in the same direction with itself when it ceases to flow. The consequence is, that the quantity of the current in either helix is not altered, but its intensity is reduced, owing to the slowness of its development. This is always the effect of any closed circuit in the neighborhood of an inducing magnet or current, on other circuits near it.

473. If the cups of the fine wire coil are joined by a wire, it will form a closed circuit around the magnet, and will impair the spark when the current in the coarse wire is interrupted, though not to so great an extent as the brass tube, since the latter offers a freer and shorter circuit for the induced current. The spark is but slightly lessened when shocks are taken from the fine wire coil, because the human body is too poor a conductor to allow of the ready flow of the secondary through it. A metallic cylinder surrounding the helices will neutralize the sparks and shocks as completely as the enclosed tube.

474. When a bar of iron is placed within the helix, a secondary is induced in it in the same manner as in the brass tube, which somewhat retards the secondary currents in the coils. Hence the greater shock obtained from a bundle of wires, where this secondary current cannot circulate. To this cause is added another — the more sudden change in the mag-

netism of the wires, when the battery current ceases, from the neutralizing influence of the similar poles of the wires on each other.

475. If the secondary current can be hindered from circulating in the brass tube, its retarding influence will be prevented. Thus, if the tube is longitudinally divided on one side, it no longer diminishes the shock or spark. With the solid iron bar, the shock and spark are increased by sawing it open longitudinally to the centre. A soft iron tube, divided like the brass tube, gives a stronger shock than the bar, but is still inferior to the bundle of wires. The two brass caps at the ends of the fine wire coil would exert a considerable neutralizing influence if they were not divided on one side, as shown in the cut. The ends of the caps are also cut through for the same reason.

476. In this instrument there are some peculiarities in the shock occasioned by the motion of the battery wire over the rasp. If it is moved slowly, distinct shocks are experienced; if the motion is quickened, the arms are much convulsed; and if it is drawn over rapidly, the succession of shocks becomes intolerably painful. This, however, can be easily regulated. The shock from the secondary coil increases within certain limits in proportion to the length and fineness of the wire of which it is composed. There is no advantage obtained by employing a very long wire, unless the battery is powerful. The shock is also lessened if a very fine wire is used, unless its length is moderate.

477. The strength of the shock depends greatly upon the extent of the surface of contact between the hands and the metallic conductors. Thus, if two wires are fixed in the cups C and D, and grasped in the hands, the shocks will be slight in comparison with those given by the handles, and still more so if the wires are held lightly in the fingers. These effects, as well as the increase of the shock by wetting the hands, are due to the comparatively low intensity of the secondary current, which causes it to be transmitted imperfectly by poor conductors. With frictional electricity it is well known that no difference in the shock is thus occasioned.

478. When the quantity of the secondary current is very small, an imperfect conductor, or a surface of limited extent, may be able to convey the whole of it, even if its intensity is not very high; in which case, the sensation and muscular contractions produced by it will not be increased, but even lessened, by any further increase in the conducting power. Thus, if the shocks are received by placing the hands in two vessels of water connected with the cups of the outer helix, and the current is rather feeble, it will produce the strongest sensation when the ends of the fingers only are immersed. With a powerful current, the shock is intolerable, whether the surface of contact with the water is large or small; in the latter case, it extends to a less distance up the arms, though it may be felt very strongly in the fingers.

479. The shocks have sufficient intensity to pass without much diminution through a circuit formed

by several persons with their hands joined, especially if their hands are moistened. Different individuals will be found to manifest remarkable differences in regard to susceptibility to the shocks; some being but slightly affected, perhaps feeling the shocks only in the hands or arms; while others will feel them as far as the shoulders or across the breast, and will experience strong muscular contractions in the arms.

480. The difference in the strength of the shock in the two arms, which has been described in the case of the magneto-electric machine (see § 442,) is exhibited more satisfactorily by the separable helices, as a rapid succession of shocks may be obtained of very nearly the same intensity. Suppose the handle connected with the positive cup of the exterior helix to be held in the right hand, and the one connected with the negative cup in the left hand. The left hand and arm will then experience the strongest sensations, and be most convulsed. In determining the positive or negative character of the cups, regard should be had only to the terminal secondary current, it being found that the initial secondary, whether induced by means of a voltaic battery or a permanent steel magnet, produces comparatively feeble physiological effects, and consequently need not, in this case, be taken into account. This singular difference in the intensity of the shocks is regarded as a purely physiological phenomenon, the greatest effect, both as respects sensation and muscular contractions, being produced by the electric current when it proceeds in the direction of the ramification of the nerves.

481. If the ends of the secondary wire are put into vessels of water, a peculiar shock may be taken by putting the fingers or hands into the vessels, so as to make a communication between them through the body. If both wires are put into a trough, at some distance apart, and two fingers of the operator are placed in the water in a line between the two wires, a shock will be felt. Here the current prefers a passage through the body to that through the water which intervenes between the fingers. The conducting power of the water may be made better than that of the human body by the addition of a sufficient quantity of common salt; in which case, little or no shock can be perceived. If the fingers are placed at right angles to the line between the wires, no shock will be felt. The trough should not be of metal, but of some poor conductor of electricity.

482. If a delicate galvanometer is connected with the ends of the fine wire coil, the needle will be deflected in opposite directions, and equally far, when the battery circuit is closed and opened. The same effect is produced when the brass tube is slipped over the iron wires. In this case, though the shock may have been prevented, the induced current still passes. The reduction in the intensity of the current, while its quantity remains unaffected, depends upon the same cause as with the flat spirals, when a metal plate is interposed. (See § 421.)

483. When a flat coil of fine wire, such as that represented at W, in Fig. 167, is passed over the in-

terior helix (the exterior one being removed), the shocks will be found strongest when the coil surrounds the middle of the helix, and to decline considerably in strength as it is either raised or depressed from this position. Now, the magnetism of the enclosed iron wires, which induces the principal part of the current, manifests itself chiefly at the ends of the bundle; it might, therefore, have been expected that the flat coil would give the strongest shock when surrounding one of these ends. The shocks from the exterior helix are also lessened when it is raised from the stand so as to enclose only the upper part of the inner helix.

484. Slight shocks may be obtained from the inner helix itself, by connecting one of the handles with the cup, A, and the other with the rasp, B. The bundle of iron wires should be within the helix. The shocks are somewhat stronger when one handle is in connection with the rasp, and the other with the battery wire which is drawn over it; in this case, the battery is included in the circuit of the secondary current.

485. The most important principles of magneto-electric and electro-dynamic induction are conveniently illustrated by the separable helices, in consequence of the facility with which the powers and uses of its several parts can be exhibited. The observations which have been made with regard to it apply equally well to the following instrument, which is a modified form.

486. Separable Helices and Electrotome.—In the instrument represented in Fig. 181, the inner

helix is connected with an electrotome, similar to that described in § 388, fixed on the same base board, in

Fig. 181.

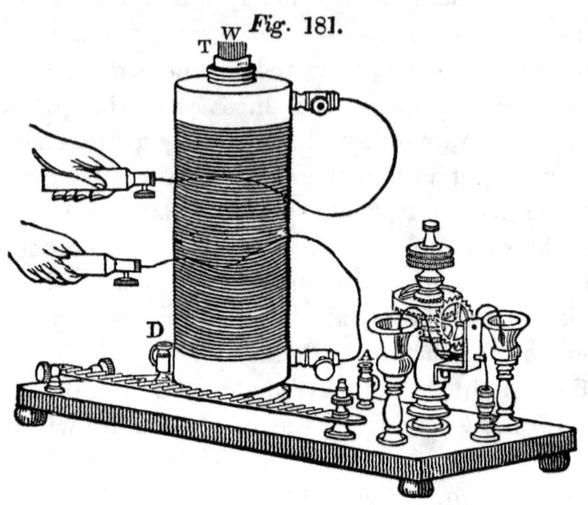

addition to the steel rasp. There are two cups, A and D, for the battery wires; these are connected, through the electrotome, with the inner helix. When the electrotome is set in motion, the curved wire dips its ends alternately into the glass cups containing mercury, and rapidly breaks the circuit. One end of the coarse wire coil is also connected with the steel rasp, so that this may be used as in the last-described instrument, when the current is not made to pass through the electrotome. At W is seen the end of the bundle of wires, and at T the brass tube, which may be slipped over them at pleasure. This instrument, and others resembling it in being provided with a mechanical contrivance for breaking the battery circuit, may be used with a very small battery, although its effects are of course more striking with a powerful one.

487. When the circuit is broken at the surface of the mercury, an intensely brilliant spark is seen, and the mercury is deflagrated, passing off in a white vapor. If the quantity of mercury is properly adjusted, the sparks occur alternately in the two cups, and in such rapid succession as to appear simultaneous. A little water or oil upon the surface of the mercury diminishes the brilliancy of the sparks, but increases the intensity of the shocks.

488. These sparks are of so short duration that moving objects appear stationary by their light. The revolving armature (Fig. 142), although rotating many hundred times a minute, appears at rest when viewed in this way; and where the sparks succeed each other rapidly, it appears to leap from place to place as their light falls on it. The revolving electro-magnet (Fig. 146), and other instruments which exhibit rapid rotation, present the same phenomena. Many optical illusions of this kind may be observed, as in moving the fingers rapidly, when their number seems increased, or rapidly turning over the leaves of a book, when they seem to leap in the same manner as the armature.

489. If the ends of the secondary wire are separated from each other at the same moment that the battery circuit is broken, a spark will be seen from the passage of the induced current. A beautiful light is produced if prepared charcoal points are

attached to the ends of the secondary wire and separated in the same way.

490. Water may be decomposed by connecting the ends of the fine wire helix with an instrument for that purpose, having very small platinum wires guarded with glass, as originally used by Wollaston. These are prepared by inserting the wires into capillary glass tubes, which are heated till the glass melts and adheres to their ends so as to cover them completely. The platinum points are then exposed by grinding away the glass. It is of course only necessary to cover those parts of the wires intended to be immersed in the fluid.

491. The extremities of the platinum wires, while the decomposition is going on, appear in a dark room, one constantly and brightly, and the other intermittingly and feebly luminous. If the apparatus for decomposition is removed out of the noise of the electrotome, rapid discharges are heard in the water, producing sharp, ticking sounds, audible at the distance of eighty or one hundred feet, and occurring at the moments when the battery circuit is broken. Decomposition is effected both by the initial and terminal secondary currents; that is to say, by the currents induced both on completing and on interrupting the battery circuit; but the ticking noise and sparks, accompanying the rapid discharges in the water, are produced only by the terminal secondary current. Both gases, hydrogen and oxygen, are given off in small quantities at each wire. The secondary current of the magneto-electric machine presents the same phenomena with the guarded points.

492. A Leyden jar, the knob of which is connected with its inside coating by a continuous wire, may be feebly charged, and slight shocks be rapidly received from it, by bringing the knob in contact with one of the cups of the outer helix, and grasping with the two hands respectively the outer coating of the jar and a handle connected with the other cup. A gold leaf electroscope is readily affected by touching its cap with a wire fixed in either cup of the exterior helix. If the contact, which should be only momentary, is made at the instant of the rupture of the primary circuit, the gold leaves exhibit a considerable divergence without the aid of a condenser. Or the knob of a Leyden jar may be touched for a moment with the wire, when it will be found to retain a feeble charge, capable of diverging the gold leaves and of giving a slight shock. The wire must be well insulated from the hand in which it is held, or the electricity will be conveyed off.

493. If the large thermo-electric battery (Fig. 41) is connected with the cups A and D, and the vibrating wire put in motion, faint sparks will be seen in the mercury cups, attended by audible snaps. Strong shocks may be obtained by grasping the handles attached to the fine wire coil, especially if both heat and cold are applied to the battery. A single thermo-electric pair, of antimony and bismuth, or of German silver and brass, connected with A and D, will give a slight shock to the tongue when heated by a spirit lamp; it will be more perceptible when the ends of two wires fixed in the cups are

made to touch the tongue than with a more extended surface of contact. This is probably due to the small quantity of the induced current. These sparks and shocks are, of course, not strictly thermo-electric, but magneto-electric.

494. When a bar of iron is contained within a horizontal helix, such as is represented in Fig. 168, where the circuit can be rapidly broken, and a small key or some tacks are applied to one end of the bar, they do not cease to be sustained, notwithstanding its magnetic attraction is intermitted every time the voltaic circuit is interrupted, since it is almost instantaneously renewed. This experiment succeeds best when the iron bar is enclosed in a brass tube within the helix, the closed circuits of the tube tending to prolong its magnetism.

495. If an iron tube of sufficient diameter to admit a long helix of fine wire within it is itself passed into a coil of coarse wire, no shocks can be obtained from the enclosed helix, even when the tube is divided longitudinally on one side, to prevent the flow of a current in its substance which might neutralize that of the fine wire. This is the converse of the fact stated in § 265, that a galvanic current passed through a coarse wire helix, enclosed in an iron tube, induces no magnetism in the tube. Let one or two iron wires be placed in a glass tube contained within a tube of iron, the whole being surrounded by the heliacal ring described in § 273. The wires are about as long as the tube, and their ends are made to extend a little beyond one of its extremities.

When the helix is connected with the battery, polarity is induced in the wires in the same direction as in the iron tube, and they are repelled until about half of their length has passed out of the tube. If the circuit is broken at this instant, the momentum which they have acquired causes them to be projected to some distance. In this case, the repulsion overpowers the axial force, which tends to draw the wires within the coil. The action is strongest when the helix surrounds that end of the tube from which they are projected.

496. Double Helix and Vibrating Electrotome.—The instrument represented in Fig. 182 is one of the most convenient foı the medical application of electricity. It is provided with a self-acting interruptor, by which shocks can be given with extreme rapidity. The double helix is secured to the base board, in a horizontal position, by two brass

Fig. 182.

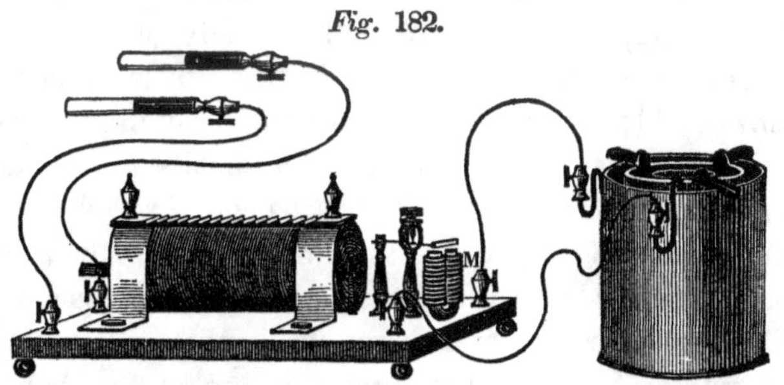

bands. The helices are insulated from each other, but are not separable. A bundle of iron wires is shown in the cut, within the inner helix. This can be removed at pleasure. The vibrating electrotome, which is fixed on the stand, is of the same construction as that represented in Fig. 176, except that the coil surrounding the electro-magnet, M, is shorter. Near the electrotome are two screw-cups for battery connections. When the battery is applied, as shown in the cut, the current traverses in succession the coarse wire helix and the coil of the electrotome. The electro-magnet is instantly charged, and attracts its armature, causing the circuit to be broken in the manner described in § 456. At the other extremity of the base board are the screw-cups belonging to the fine wire helix. With these the handles for shocks are connected. With all the magneto-electric instruments, the battery connections must be made by stout copper wires. For attaching the handles to the cups of the outer coil, rather fine wire is more convenient.

497. With a battery of even moderate power, the shocks may be made to follow each other with exceeding rapidity. When their strength is lessened considerably by removing nearly all the iron wires from the helices, instead of distinct shocks, a peculiar sensation of numbness is experienced, extending a greater or less distance up the arms, and attended by loss of power over the muscles as far as it reaches. The shocks are never so powerful with a self-acting interruptor as when the circuit is broken mechanically, since the battery current is obliged to maintain the motion of the interruptor as well as to traverse a circuit of greater length.

498. There is a steel rasp above the helices, by means of which shocks are given without using the electrotome. For this purpose, the battery wire is removed from one of the screw-cups near M, and drawn over the rasp. It will be found by trial from which of the cups it is necessary to remove the wire. If the hollow of the helix is filled with iron wires, bright sparks will be seen as the battery wire leaves each tooth of the rasp, and strong shocks will be felt when one of the handles is grasped in each hand. When the iron wires are withdrawn, the spark becomes faint and the shock feeble.

499. The strength of the shock may be regulated by varying the number of iron wires which are placed within the helix, or the distance to which the bundle is allowed to enter. The addition of a single wire produces a perceptible increase in the shock, especially when only a few are already within. By wetting the hands or other parts to which the handles are applied, especially with salt water, the shock is still stronger. It may, on the contrary, be lessened in some degree by diminishing the extent of contact between the handles and the surface of the body. If, however, the current is powerful and the contact too slight, a disagreeable burning sensation will be experienced at the part touched by the metal.

500. The shocks may be passed through any portion of the body, by placing the handles so as to include that part in the path of the secondary current; their intensity is greater when the handles are

near each other. The influence does not extend beyond the direct course of the current, unless the shocks are severe. When, however, one of the handles is placed directly over a large and tolerably superficial nerve, the shock will be felt not only in the parts intervening between the handles, but through those to which the ramifications of the nerve are distributed. Thus, if one handle is held in the right hand, and the other pressed upon the inside of the left arm over the median nerve, the sensation will be experienced even to the ends of the fingers, attended by convulsive movements of their muscles. This is, unquestionably, a physiological phenomenon, and not a consequence of the flow of the current below the position of the handle. The difference in the intensity of the shock in the two arms, described in § 480, may be observed with this instrument.

501. Compound Magnet and Electrotome.—In Fig. 183 a double helix is seen attached to the base board by two brass bands. It is placed in a horizontal position, and within it a bundle of soft iron wires is permanently fixed. There are two screw-cups for the battery connections at one end of the stand; one of these is connected with the band which sustains the glass mercury cup, C. To the second screw-cup is soldered one end of the coarse wire coil, the other extremity of which is connected with the band upon which the brass cup, B, also intended to hold mercury, is fixed. A bent wire, W, moving on a horizontal axis, dips its ends into the two mercury

cups. On the opposite side of the axis is attached a curved iron rod, R, the lower extremity of which

Fig. 183.

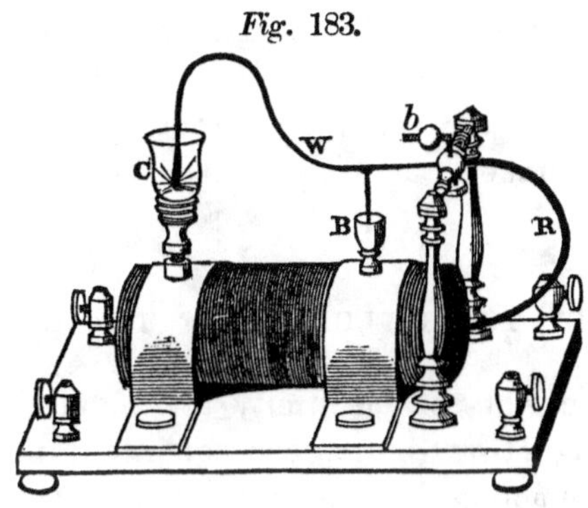

approaches nearly the end of the enclosed bundle of iron wires.

502. When the connections are made with the battery, the current traverses the wire, W, and the inner helix, causing the iron wires to become magnetic. They now attract the end of the iron rod, R; the motion of the rod raises the bent wire out of the mercury in the cup C, and breaks the circuit. This destroys the magnetism of the iron wires, and R is no longer attracted. The wire, W, then falls back by its own weight, and the circuit is renewed. A thin slip of brass is brazed to the end of R, to prevent it from being retained by the electro-magnet after the rupture of the circuit.

503. In this manner a rapid vibration of the wire is produced, and brilliant sparks and deflagration of the mercury take place in the cup C. The proper

26 *

balance of the vibrating apparatus is insured by the brass ball, *b*, which moves on a screw cut in a bent wire above the axis. The ends of the fine wire coil are connected with the two screw-cups at the opposite end of the base board to those for connection with the battery. From these cups, the shocks are taken.

III. BY THE INFLUENCE OF THE EARTH

504. Currents of electricity are induced by terrestrial magnetism; but, in consequence of the feebleness of the action, it is not easy to render it sensible by the aid of wire coils alone. Deflections may, however, be obtained by connecting with a very delicate galvanometer a helix of coarse wire, such as is represented in Fig. 113, or a flat spiral, Fig. 112, and, having placed its axis in the line of the *dip*, suddenly inverting it.

505. Stronger deflections are produced by causing a helix to revolve rapidly, as in the instrument represented in Fig. 184. The coil which is hollow, moves in a vertical plane, and its shaft is provided with a pole-changer, to the segments of which the extremities of the wire are soldered. The springs pressing on these segments convey the currents to the screw-cups on the base board. When the cups are connected with those of a delicate galvanometer, and the instrument placed in such a direction that the helix shall move in the magnetic meridian, it is made to revolve rapidly by means of a multiplying wheel. As each end of the helix approaches and recedes from the line of the dip, opposite currents

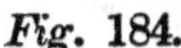

Fig. 184.

are induced in the wire, their direction changing as the helix passes this point.

506. These alternating currents of electricity are turned into one course by the pole-changer, the segments of which may be so arranged as to pass from one spring to the other when the helix is vertical, this being sufficiently near the line of the dip, except in low latitudes. The galvanometer needle is steadily deflected as long as the motion is maintained uniformly. By reversing the revolution of the helix, a deflection in the opposite direction takes place.

507. A still more powerful effect is produced by fixing an iron bar within the hollow helix. Let the instrument be now placed in the plane of the magnetic meridian, and the cups connected with a galvanometer. When the coil and bar are made to re-

volve, each end of the iron becomes alternately a north and a south pole, the magnetism being induced in it by the earth, as explained in § 354. By the changes in the magnetic state of the iron, electric currents are induced in the surrounding coil. These currents change their direction twice during each revolution, but they are brought into one course by the pole-changer on the shaft of the rotating coil, and the needle of the galvanometer is strongly and steadily deflected. With this instrument, the current is somewhat augmented by the feeble one excited in the wire coil by the direct magneto-electric induction of the earth.

508. In this, and in all other instances where electricity is produced by motion, and motion reciprocally by electricity, the motion must be the reverse of that which would result from a galvanic current flowing in a certain direction, in order that the induced current may have the same direction. An opposite current is excited by producing mechanically the same motion as that obtained with the battery.

INDEX.

CATALOGUE

Figure.		Price.
17.	Grove's Battery, (single pair,)	$2,00 to 2,50
18.	Grove's Battery, in box, (per pair,)	1,50—2,00 & 2,50
20.	Large Trough Battery, (100 pairs,)	85,00
21.	Grasshopper Battery,	,25 to ,50
22.	Frog, or Leech Battery,	50 to ,75
25.	Double Forceps, on stand,	2,00 to 5,00
26.	Powder Cup, of glass,	,25
	Powder Cup, of brass,	,50
27.	Voltaic Gas Pistol,	2,50
28.	Decomposing Cell, with one gas tube,	2,50
	Decomposing Cell, with two gas tubes,	3,50
29.	Cell for delicate decompositions,	2,00 to 3,00
30.	Divided Tumbler, for decompositions,	2,00 to 3,00
31.	U-Tube, on stand,	3,50
32.	Smee's Battery, (improved,)	3,00 to 5,00
34.	Smee's Battery, for gilding and silvering, (per pair,)	2,00
35.	Smee's Battery, for gilding, &c., (improved,)	1,50 to 2,00
38.	Thermo-Electric Pair,	,25
39.	Delicate Galvanometer,	3,00 to 5,00
40.	Thermo-Electric Series, (10 pairs,)	2,00
41.	Thermo-Electric Battery, (60 pairs,)	25,00
42.	Cold and Heat Instrument,	3,00
52.	Astatic Needle, on stand,	1,50 to 2,00
53.	Simple Galvanometer,	2,50 to 3,50
54.	Horizontal Galvanometer,	6,00
55.	Upright Galvanometer,	5,00
56.	Magnet revolving round Conductor,	6,00
57.	Magnet revolving on its Axis,	6,00
58.	Revolving Wire Frames,	8,00
59.	Revolving Cylinders,	8,00
60.	Rotating Batteries,	8,00
61.	Vibrating Wire, (with magnet,)	5,00
62.	Vibrating Wire, with fixed magnet,	6,00
63.	Gold Leaf Galvanoscope,	5,00 to 7,00
64.	Revolving Spur-Wheel,	5,00 to 8,00
65.	De la Rive's Ring, in small glass cup,	1,00
66	Vibrating Coil Engine,	8,00
67.	Reciprocating Coil Engine,	10,00
68	Reciprocating Magnet Engine,	10,00

Figure		Price.
69.	Revolving Coil, ...	$6,00
71.	Revolving Rectangle, ...	6,00
72.	Revolving Galvanometer Needle, ...	6,00 to 7,00
73.	Revolving Coil and Magnet, ...	8,00 to 10,00
74.	Thermo-Electric Revolving Arch, ...	2,00
75.	Thermo-Electric Arch, on U-magnet, ...	4,00
76.	Thermo-Electric Revolving Frames, ...	6,00
77.	Thermo-Electric Arch, between poles of U-magnet,	4,00
78.	Double Thermo-Electric Revolving Arch, ...	6,00
79.	Magnetic Needle, (on brass stand,)	75 to 1,50
80.	Large Dipping Needle, on universal joint, ...	4,00 to 6,00
82.	Bar Magnets, in frame, (to illustrate Earth's magnetism,)	6,00
84.	Globe, with Coil and two Needles, ...	4,00
87.	Star-shaped Iron Plate, ...	,75
89.	U-Magnet and Rolling Wires, ...	1,00 to 2,00
90.	Y-Armature, ...	,30
92.	Rolling Armature, with U-magnet, ...	2,00
93.	Magnet Armature and Roller, ...	1,50
94.	Short Rolling Wire, with U-magnet, ...	1,00 to 1,50
99	Iron Wire, sealed up in glass tube, (with magnet,)	,75
100.	Instrument for measuring Attractive Force, ...	15,00
101.	Bar Magnet, for fracturing, each, ...	,12
107.	Helix, on stand, (with iron bar,)	2,00 to 2,50
108.	Heliacal Ring and Semicircles, (ball and socket joints,)	5,00
109.	Heliacal Ring and Semicircles, ...	2,50 to 4,00
110.	Grove's Battery, with covered top, ...	6,00 to 8,00
111.	Heliacal Ring and Double Cylinders, ...	4,00 to 6,00
112.	Flat Spiral, per lb., ...	,75 to 1,00
113.	Magnetizing Helix, per lb., ...	1,25 to 1,50
114.	Axial Bell Engine, ...	8,00
115.	Axial Revolving Bar, ...	6,00 to 8,00
116.	Inclined Revolving Bar, ...	5,00 to 6,00
117.	Inclined Bar revolving round Conductor, ...	6,00
118.	Axial Revolving Circle, ...	10 to 12,00
119.	Vibrating Helix, ...	4,00
120.	Double Helix and U-shaped Bar, ...	1,50
121	Double Axial Bell Engine, ...	8,00
122.	Upright Axial Engine, ...	15,00
123.	Horizontal Axial Engine, ...	18,00

Figure.		Price
124.	Double Beam Axial Engine,	$ 15,00 to 18,00
125.	Single Beam Axial Engine,	15,00
126.	Small Electro-Magnet,	,75 to 1,00
127.	Electro-Magnet, in frame,	8,00 to 12,00
128.	Electro-Magnet, in case,	10,00 to 25,00
129.	Electro-Magnet, with three poles,	3,00
130.	Steel Bar, of the U form, for charging,	,25 to 50
131.	Magnetometer,	12,00
132.	Axial Magnetometer,	12,00 to 15,00
133.	Electro-Magnetic Telegraph,	5,00
134.	Telegraph, with Clock-work,	25,00 to 35,00
135.	Signal Key,	2,50
136.	Axial Receiving Magnet,	20,00
137.	Axial Telegraph,	6,00 to 10,00
138.	Axial Telegraph, with Clock-work,	25,00 to 35,00
139.	Axial Telegraph, with Engine,	35,00
140.	Reciprocating Armature Engine,	10,00 to 12,00
141.	Horizontal Reciprocating Engine,	18,00 to 20,00
142.	Revolving Armature,	3,50
143.	Revolving Armature Engine,	6,00
144.	Vibrating Armature Engine,	10,00
145.	Vibrating Armature Engine,	8,00
146.	Revolving Electro-Magnet,	5,00
147.	Revolving Bell Engine,	10,00 to 12,00
148.	Registering Revolving Magnet,	6,00 to 8,00
149.	Electro-Magnet, revolving by the Earth's action,	8,00
150.	Electro-Magnet, revolving by the Earth's action,	7,00
151.	Electro-Magnet, revolving within a coil,	8,00
152.	Double Revolving Magnet,	7,00 to 8,00
153.	Revolving Wheels, with electro-magnet,	8,00
154.	Revolving Wheels,	8,00 to 10,00
155.	Electro-Magnet, revolving on its axis,	7,00
156.	Iron Bar, for showing magnetic induction of Earth,	,75
159.	Attracting and Repelling Wires,	3,50
160.	Contracting Helix,	3,50
162.	Electro-Dynamic Revolving Coil,	6,00
163.	Electro-Dynamic Revolving Rectangle,	7,00
164	Instrument for showing the revolution of Mercury,	4,00
	Same Instrument, with iron cup,	5,00

Figure.		Price.
165.	Clock-work Electrotome, ...	$ 6,00 to 7,00
167.	Fine Wire Coil, per lb., ...	1,50
168.	Double Helix and Electrotome,..............	20,00
172.	Galvanometer, Astatic Needle, ...	5,00
173.	Magneto-Electric Machine, ...	30,00 to 35,00
174.	Improved Magneto-Electric Machine, ...	40,00 to 45,00
175.	Magneto-Electric Machine, with Vibrating Electrotome,	45,00
176.	Vibrating Electrotome,......................	4,00 to 5,00
177.	Magneto-Electric Machine, for quantity, ...	40,00
179.	Separable Helices, with handles, ...	10,00 to 12,00
181.	Separable Helices and Electrotome,..........	20,00
	Apparatus for showing the Suspension in Air of an Iron Rod by Repulsion,...............	3,00
182.	Double Helix and Vibrating Electrotome, with Battery and Handles,.....................	10,00 to 12,00
183.	Compound Magnet and Electrotome,.........	8,00 to 10,00
184.	Apparatus for showing Induction of the Earth,	12,00

INSTRUMENTS IN THE MEDICAL APPLICATION OF ELEC TRICITY.—SECOND EDITION.

Figure.		Price.
1.	Galvanic Battery, (according to size,)	$ 2, 5, 7,00
2.	Electro-Magnetic Apparatus, with Battery and Handles,...............................	8,00 to 10,00
3.	Electro-Magnetic Apparatus and Revolving Armature with Battery,..................	15,00
4.	Electro-Magnetic Apparatus and Vibrating Armature, ...	10,00 to 12,00
5.	Electro-Magnetic Box Apparatus, (complete,).	12,00
6.	Magneto-Electric Machine,..................	30,00 to 35,00
7.	Improved Magneto-Electric Machine, ...	25,00 to 30,00
8.	Box Magneto-Electric Machine, ...	30,00 to 35,00
	Covered Copper and Iron Wire, per lb.	
	Fine Platinum Wire for Experiments.	

www.ingramcontent.com/pod-product-compliance
Lightning Source LLC
LaVergne TN
LVHW020215110826
845151LV00003B/721
9781425533885